Mathematics For Everyone

is the keynote of this delightful and instructive book which brings the fascinating world of numbers within easy grasp of any reader.

Ever since the time of the ancient Egyptians, man has seen the world of numbers as one of magic, full of fantasy and enchantment. Professor Tocquet's book opens up and explores this world, giving the reader a unique insight into the strange properties of numbers and the mysteries of mathematical phenomena.

Revealed here in clear, everyday language are the basic steps to the mastery of mathematical puzzles, the art of rapid mental calculation, and a variety of ingenious numerical devices heretofore known only to magicians and experienced mathematicians.

THE MAGIC OF NUMBERS

by Robert Tocquet

Published in France as:
2 + 2 = 4

Translated by
DENIS WEAVER

A Premier Book

Fawcett Publications, Inc., Greenwich, Conn.
Member of American Book Publishers Council, Inc.

A Premier Book published by arrangement with
A. S. Barnes & Company, Inc.

First published in France in 1957 by Pierre Amiot as
"2 + 2 = 4"

First American edition, 1961.

First Premier printing, August 1962

Premier Books are published by Fawcett World Library,
67 West 44th Street, New York 36, New York.
Printed in the United States of America.

Contents

While the facts given in this book are sound, the author's intention is to amuse rather than instruct. For this reason "textbook" terminology has been avoided wherever possible.

CHAPTER I

Yesterday and Today

LIGHTNING CALCULATORS, especially when illiterate, have drawn the attention of the public in all ages by their extraordinary abilities. They can solve in their heads, sometimes instantaneously and without apparent effort, problems often so complicated that most of us, even mathematicians accustomed to juggle with figures, could solve them only with pencil and paper and over a much longer time, without being sure even then of succeeding. Some of them, too, when they have been set a problem, can talk freely with bystanders, discussing subjects completely foreign to the question they are dealing with, and then suddenly give the required solution, as if a cerebral mechanism had been working within them without their knowledge.

As a general rule, and this is a fact which should be emphasized immediately, the lightning calculators, apart from the faculty they have of handling figures with exceptional virtuosity, are of below average intelligence; sometimes they are even mentally retarded. Thus, Colburn was always at the bottom of his class, Buxton could not even write his name and Inaudi did not learn to read or write until he was over twenty years of age. There are certain exceptions, however, to this rule, for some have been known who have educated themselves normally and there have even been geniuses who were phenomenal calculators: Ampère, Arago, Georges Bidder, Whately and Gauss being examples.

Lightning calculators of the past

Among calculating prodigies who were otherwise backward or who had very little education, let us recall those who had the greatest renown in the past before examining present-day calculators in greater detail.

DWG

The Greek writer Julian mentions a certain Nikomachos, who lived at Gerasa in Palestine in the second century of our era, and who found solutions to difficult problems very rapidly.

Balthasar of Monconys, in an account of his third journey in Italy, records that in 1664 Mathieu le Coq, then aged eight and unable to read or write, had been performing advanced arithmetical operations, such as multiplications with five or six figures and extractions of square and cube roots, for some two years previously.

Thomas Fuller, nicknamed the Virginian Calculator, or the Negro Calculator, was almost totally ignorant. A slave in Virginia in the middle of the eighteenth century, he could neither read nor write and he died at the age of eighty without ever having learned to do so. Scripture records the following story about him in the *American Journal of Psychology:*

"When Fuller was about seventy years old, two gentlemen of Pennsylvania, William Hartshorne and Samuel Coates, both men worthy of confidence, heard of the calculator and had the curiosity to have him brought before them and put to him the following problems: First, how many seconds are there in a year and a half? Fuller replied in two minutes that there are 47,340,000 seconds. Secondly, how many seconds has a man lived who is aged seventy years, seventeen days and twelve hours? Fuller replied at the end of a minute and a half: 2,210,800,800. One of the gentlemen who examined him had taken the trouble to do the calculation on paper and told Fuller he was wrong and that the number of seconds was less. But Fuller pointed out promptly that this difference in the two results had to do with leap years."

Jedediah Buxton, who was born at Elmeton, near Chesterfield, lived in England from 1702 to 1762. A poor laborer, totally illiterate to the point that he was unable to scribble his own name, of below average intelligence, Buxton had the greatest difficulty in learning the multiplication table. This was in fact the only education he received. On the other hand, like most lightning calculators, he had a highly developed memory for figures. For example he knew the exact number of seconds in a day and in a year. Possessed of a positive mania for arithmetic, "he could perceive only figures, and pretexts for mental operations," writes Alfred Binet, "his mind being completely closed to anything else." When he came to London he was taken to Drury Lane Theatre to see Richard III played by Garrick. When he was asked afterwards if the performance had pleased him it appeared that he had found in it only an opportunity to perform calculations; during the dances he had fixed his attention on the number of steps executed; there were 5,202; he had also counted the number of words the actors had spoken: 12,445; he had also remembered the number of words spoken by Garrick, and all this was found to be accurate. In his calculations he reduced all lengths to an eccentric standard, the thickness of a hair, which he had probably fixed arbitrarily.[1]

He was not merely a mental calculator of great capacity; he had also a very accurate eye, a sort of divinatory power for areas. When he went through a field or a simple piece of land, he could, so it is said, give the area as exactly as if he had measured it with a surveying chain. In this way he determined the extent of the Elmeton estate in acres, and for his personal satisfaction, worked out the result in square inches and in squares having the thickness of a hair.

Zerah Colburn, who was born in the state of Vermont, in 1804, began calculating without knowing how to read or write. One day his father noticed his unusual aptitude by accident. The child was repeating aloud the products of the multiplication table without, it seems, ever having had the table at his disposal. Astonished, Mr. Colburn asked him:

[1] A few years ago Dr. Ginestous brought to the Bordeaux Anatomical Society a young man who in a sense behaved in somewhat the same way as Buxton. He continuously counted in his head all the letters contained in the sentences he read, wrote, spoke, heard or thought. Whether he was reading a newspaper, writing something down, or chatting with a friend or a stranger, he counted everything without this fantastic labor occasioning him the slightest fatigue, causing him any physiological difficulty or hindering him in his profession. From his childhood he counted all the letters in his lesson books and those in books he looked through by chance, and in the lines of the Aeneid or the Iliad which he had to learn by heart. This strange mental activity ceased at night and he did not dream. But it recommenced on waking. When he was not being spoken to or was not reading he made up sentences which he ceaselessly counted.

This subject displayed one other peculiarity. He had the phenomenon of colored vision. The months and the days of the week presented themselves to him in a color lighter or darker according to the passage of time towards the last month of the year or the last day of the week. January and Monday appeared to him as perfectly white; the other months and the other days of the week were slightly tinged with gray, which darkened gradually to attain complete blackness in December and on Sunday.

"How much is 13 times 97?" and the child replied at once: "1,261." He was then six years old.

Mr. Colburn saw in this gift for calculation a means of earning money and had the idea of exhibiting his son. Zerah Colburn thus became the first of the professional lightning calculators. He performed at Boston, and then, at the age of ten, came to London, and to Paris where, apparently, his performances had no great success. When an attempt was made to get him to reveal the secret of his mysterious gift, he invariably replied: "It is God who has put these things into my head and I would not know how to put them into yours."

Thanks to the support and patronage of Washington Irving he was admitted as a pupil at the Lycée Napoléon in Paris, then at the College of Westminster, but it was found that apart from mental calculation his mind was closed to all normal forms of instruction. He lost his faculty to calculate, without any apparent reason, at the age of twenty.

It is worthy of note that Colburn had one curious physical peculiarity; a certain sign of degeneracy: he had an extra finger on each hand and an extra toe on each foot.

Zacharias Dase, born in Germany in 1824, distinguished himself from the majority of lightning calculators by the fact that he placed his ability at the service of science. He calculated the natural logarithms of the numbers from 1 to 100,500 and the table of factors and prime numbers from the seventh to the eighth million. But he was never able to learn conventional mathematics, despite the efforts of the eminent teachers who interested themselves in him; in addition he showed no intelligence in anything which did not have to do with figures or numbers. Both his faculty for counting and his memory were prodigious: the astronomer Gauss made him multiply *mentally* one by the other, numbers each composed of *a hundred figures*. One is literally stupefied to realize pen in hand what an enormous array of figures such an operation entails. Yet Schumacher reports that Dase multiplied two numbers of 8 figures each in fifty-four seconds and two numbers of 20 figures each in six minutes. He had also a great rapidity in perception and a visual memory for estimating the number of objects contained in a group, for instance the number of books in a library.

Vito Mangiamele was a little ten-year-old Sicilian shepherd boy when he came to Paris to be examined by Arago in 1837. Though he had had no education he had worked out for himself calculating processes which have never been properly explained. At the Académie des Sciences where he was introduced he was asked, for example:

"What is the cube root of 3,796,416?"

At the end of a minute he replied: "156," which was correct. It took him no longer to solve the two equations:

$$x^3 + 5x^2 - 42x - 40 = 0 \text{ and}$$
$$x^5 - 4x - 16{,}799 = 0$$

Henri Mondeux became very famous. Born in 1826 at Neuvy-le-Roi, near Tours, he was the son of a poor woodcutter. At the age of seven, when he could neither read nor write, he used to amuse himself by doing involved calculations while guarding his sheep. Being ignorant of figures he counted with little pebbles arranged in different ways. A Tours schoolmaster, M. Jacoby, who had been told of this, tried to teach him but without any success.

Ultimately he was taken to Paris and presented to the Académie des Sciences. An examining committee was set up which included Arago, Serres, Sturm, Liouville and Gauchy.

It was speedily recognized that the child had an abnormal faculty for mental calculation, a prodigious memory for numbers, but an almost total absence of memory for the names of places or people, or of objects which did not interest him. It was observed also that while solving a problem he could engage in completely unrelated activities. Finally the committee published some of the calculating processes he had invented. These are given in the report by Gauchy from which the following passages are taken:

"In many instances Henri Mondeux invents methods, which are sometimes remarkable, for solving a number of questions which are ordinarily treated by algebra, and determines in his own fashion the exact or approximate values of whole numbers or fractions which must fulfil required conditions. When he has to multiply whole numbers by one another Henri Mondeux frequently splits these numbers into groups of two figures. He has succeeded in finding out for himself that the operation becomes simpler in cases where the factors are equal, and the rules he then uses to arrive at the product or the power required are precisely those which would yield the formula known as the binomial theorem. Guided by these rules he was able to give at the very instant he was asked for them, the squares and cubes of a multitude of numbers, for example the square of 1,204 or the cube of 1,006. As he knows pretty well by heart the squares of all the whole numbers below 100, the splitting of large numbers into groups of two figures enables him to obtain their squares more easily. Thus he was able to find the square of 755 almost immediately while in the presence of the Académie.

"Henri has succeeded unaided in discovering the established method for finding the sum of an arithmetical progression. Among the rules which he has invented for solving different problems are several which can be inferred from certain algebraical formulae. The method he has worked out for calculating the sum of the cubes, and the fourth and fifth powers of natural numbers, can be cited as an example. To solve two simultaneous equations of the first degree Henri has made use of a device which is worth mention. He has sought first for the difference between the unknown quantities and to do this he has subtracted one equation from the other, after having multiplied the first by the ratio which exists between the sums formed successively by adding the one and the other to the coefficients of the two unknowns.

"Where it is a case of solving not simultaneous equations of the first degree but a single equation of a degree higher than the first, Henri habitually uses a method which we will explain by an example. We put a problem to him of which these are the terms: Find a number such that its cube, increased by 84, furnishes a sum equal to the product of this number when multiplied by 37. Henri gave as a solution the numbers 3 and 4. To obtain them he began by transforming the equation which it was required to solve by dividing the two numbers by the number sought. The question was then reduced to the following: find a number such that its square, increased by the quotient obtained on dividing 84 by this number, gives 37. By the aid of this tranformation Henri Mondeux was able to recognize immediately that the number sought was inferior to the square root of 37, by consequence to 6; and very soon a few simple trials led him to the two numbers we have indicated.

"Even questions of indefinite analysis are not beyond the reach of Henri Mondeux. One of us asked him for two squares of which the difference was 133. As solution he gave immediately the system of the numbers 66 and 67. It was insisted that a simpler solution be obtained. After a moment of reflection he indicated the numbers of 6 and 13.

"This is the manner in which Henri proceeded to reach

the two solutions. The difference between the squares of the numbers sought exceeds the square of their difference by a quantity equal to twice this difference multiplied by the smaller. The question put can thus be reduced to the following: subtract from the number 133 a square such that the remainder will be divisible by double the root. If a trial is made one after the other of the squares 1, 4, 9, 16, 25, 36, 49 . . . it will be recognized that among these squares 1 and 49 are the only ones which satisfy the new question. By subtracting these from 133 and dividing the remainders 132 and 84 by the doubled roots, that is by 2 and 14, the quotients 66 and 6 are obtained, each of which responds to one of the solutions given by Henri Mondeux. It is perceived that in following the procedure we have described Henri did not first arrive at that of the two solutions which seems to us the simpler, but that which offers the squares whose roots are nearest to each other."

Gauchy ends his report by expressing the hope that Henri Mondeux would distinguish himself one day in a career of science, but despite this optimistic hope this lightning calculator died in obscurity.

Contemporary Lightning Calculators: Jacques Inaudi

We come now to the best known and most popular lightning calculator of our own time: Jacques Inaudi.

Jacques Inaudi was born of a very poor family, at Onorato, in Piedmont in 1867. He was a shepherd boy, nearly six years old, when he was seized by a passion for figures. While guarding his flocks, he practiced working out numbers in his head, so that by the age of seven he was already able to do mental multiplication involving five-figure numbers. And he did not know the multiplication table! After his mother died he left the neighborhood and set out to wander with his brother, exhibiting a tame squirrel. The brother played the organ, while Jacques displayed the squirrel and collected the money.

He talked to bystanders they met about mental calculations, of which they naturally understood nothing. At the markets he helped the peasants do their accounts: "As a matter of fact," he writes, "I was very surprised that these men, who were generally fairly astute, should not have known the totals of their accounts, which I knew myself, almost instantaneously, just by listening to them; this gave me the courage one day to take part in an argument about the settlement of an account between two peasants who were about to come to blows, and whom I was able to calm down after demonstrating to them that they were both wrong; this altercation naturally attracted a crowd, astonished that a little chap like me should know how to count better than the grown-ups. Those who were good at figures put various questions to me to which I replied correctly and very easily, still remaining amazed that anyone should be ignorant of answers which seemed so natural to me. Afterwards the peasants came to seek me out whenever a difficulty arose."

Soon Inaudi made his appearance in cafés where he was noticed by a commercial traveller, a M. Dombey, who became his impresario and took him on tours in the provinces and then to Paris. There he attracted the attention of Camille Flammarion, who devoted some articles to him in various scientific journals. The well known anthropologist, Paul Broca, examined him in his turn and prepared a short report on his case. Broca noted that Jacques Inaudi's head was very large and irregularly shaped. The young prodigy afterwards gave performances in the Salle des Capucines in Paris, and later at the Robert Houdin Theatre. By this time he was aged thirteen.

It was 1892 when he returned to Paris; he had learned meanwhile to read and write and his intelligence had developed somewhat. He was now a young man of 24; short and stocky, with a strong head, eyes half closed, the facial angle very developed, almost straight. According to Binet, he had a gentle and modest disposition, spoke little and was rather reserved. His education was not yet extensive; hence subjects for conversation remained limited.

His impresario, who was then M. Thorcey, introduced him to the Académie des Sciences, which formed a committee to study the calculator. It included MM. Darboux, Poincaré, Tisserant and Charcot. Alfred Binet became a member later. After numerous tests the committee gave its conclusions, via the pen of M. Darboux.

"It should be pointed out first of all," the well known mathematician wrote, "that the results of which we were the witnesses depended above all upon a prodigious memory. At the end of a session attended by our students, Inaudi repeated a series of numbers comprising more than 400 figures. At one of our meetings we gave Inaudi a number of 22 figures. He was able to repeat it eight days after, although we had not warned him that he would be asked for it again. A second point, which seems to me of the highest importance, has been neglected by the majority of those who examined him. They have analyzed with great care the processes, certainly very simple in themselves, which Inaudi uses for various operations; but they have not noted one very evident fact: that these methods have all been invented by the calculator himself, and *that they are wholly original.* And, what is also interesting is that these rules differ from those which are taught everywhere in Europe, though some of them resemble in certain respects those which are followed for example, among the Hindus. This will be evident from the following:

Addition.—Inaudi easily adds 6 numbers of 4 or 5 figures; but he proceeds successively, adding the first two, then their sum to the next and so on. He *always starts adding from the left,* as the Hindus do today, instead of from the right as we do.

Subtraction.—This is one of Inaudi's triumphs. He subtracts easily, one from the other, two numbers consisting of a score of figures, *beginning always from the left.*

Multiplication.—The methods used are all elementary, but they require the memory of an Inaudi. For example, to multiply 834 by 36 he makes the following breakdown:

$$\begin{aligned} 800 \times 30 &= 24{,}000 \\ 800 \times 6 &= 4{,}800 \\ 30 \times 36 &= 1{,}080 \\ 4 \times 36 &= 144 \\ \hline \text{Total} & \quad 30{,}024 \end{aligned}$$

In all these partial multiplications, none of the factors ever has more than one significant figure. Despite this Inaudi knows and uses the property of the factor 25; he knows that to multiply by this number it is sufficient to take the quarter of the centuple. For example, for the square of 27, he will make the following breakdown:

$$\begin{aligned} 25 \times 27 &= 675 \\ 2 \times 27 &= 54 \\ \hline \text{Total} & \quad 729 \end{aligned}$$

Sometimes he uses partial products bearing the index sign − (minus). For example, for the cube of 27, that is the product of 729 multiplied by 27, he makes this breakdown:

$$(700 \times 20;\ 700 \times 7;\ 30 \times 20;\ 30 \times 7) - 27$$
$$\text{or } 730 \times 27 = 19{,}710 - 27 = 19{,}683.$$

Division.—Here Inaudi fundamentally follows the ordinary rule, which reduces the division to a subtraction, but sometimes uses simplifications made possible by his memory, to which we must always return.

Raising to powers.—For raising numbers to powers, Inaudi knows and applies the rule relative to the square of a sum. For example, for the square of 234,567 he uses the breakdown:

$$234{,}000^2 + 2 \times 234{,}000 \times 567 + 567^2.$$

Extraction of roots.—Here no rules are followed; there is only a simple trial and error. For example, to find a root which is 14,672, Inaudi would have tried 14,000 and 15,000, then 14,600, then 14,650, 14,660, 14,670 . . . and each time, the power of the number tried would have been subtracted from the higher number.[1]

"Inaudi also solves difficult questions in arithmetic and algebra to which the answer is furnished by whole numbers."

Here are some of the problems which are not mentioned in the report by M. Darboux:

1. Find the number whose square root and cube root differ by 18. Answer: 729, given in one minute fifty-seven seconds. (*Revue Scientifique.*)

2. Find a number of two figures such that the difference between four times the first figure and three times the second equals 7, and which, reversed, reduces the number

[1] In 1948, when he was eighty-one years of age, Inaudi discovered a general law which he expressed in these terms:

"To find, for instance, the sum of the 25 first cubes, I multiply 25 by 26, obtaining 650. I divide the result by 2, which gives 325, and I raise it to its square, which gives the number required: 105,625." Thus empirically Inaudi found the formula which gives the sum of the cubes of the *n* first whole numbers and which is expressed:

$$S = \left(\frac{n\,(n + I)}{2}\right)^2$$

Henri Mondeux had also discovered this process.

by 18. Negative solution found at the end of two minutes.

3. Find a number of four figures of which the sum is 25, given that the sum of the figures of the hundreds and thousands is equal to the figure of the tens, that the sum of the figures of the tens and the thousands is equal to the figure of the units and such that, if the number is reversed, it increases by 8,082.

Answer: Since the number increases by 8,082 when it is reversed, therefore the figure of the thousands should be 1 and the figure of the units 9; I therefore take away 9, which is the figure of the units, from 25; I have 16 left for the three other figures. Next, the figure of the thousands plus that of the hundreds is equal to that of the tens; the figure of the tens must necessarily be the half of 16, that is 8. Three of the figures being known, it is sufficient to take them from 25 to have the hundreds, 7, and to find that the number required is 1,789. (*Revue Scientifique.*)

4. The sum of three numbers is 43 and that of their cubes 17,299. Answer 25, 11, 7.

5. Find a number of four figures of which the sum of the figures shall be 16, given that the 3rd is twice the 1st, that the 4th equals three times the 1st plus the 3rd. This number reversed increases by 3,456. Answer: 1,825.

6. The distance from Paris to Marseilles is 863 kilometers. A train starts from Paris at 8:15 a.m. for Marseilles at a speed of 39 kilometers an hour. Another train starts from Marseilles for Paris at 10:30 a.m. at a speed of 46 kilometers 500 meters per hour. Find at what distance from the two cities the trains will meet.

Answer: The trains will meet at 7 hr. 31 min. 13 and $\frac{4}{6}$ secs. p.m., at 419 kilometers 451 meters and 80 centimeters from Marseilles, and at 344 kilometers 548 meters and 20 centimeters from Paris.

7. M. Laurent, examiner at the École Polytechnique, having told Alfred Binet that the calculator Winckler was able to break down a number into four squares, the test was tried with Inaudi.

Alfred Binet proposed the number 13,411.

In three minutes Inaudi gave the four following numbers:

115, of which the square is 13,225; 13, of which the square is 169; 4, of which the square is 16; 1, of which the square is 1. Total of the four squares: 13,411.

One minute later the calculator found another solution:

113, of which the square is 12,769; 25, of which the

square is 625; 4, of which the square is 16; 1, of which the square is 1. Total of the four squares: 13,411.

Finally, some time afterwards (the exact time has not been specified), Inaudi produced a third solution:

113, of which the square is 12,769; 23, of which the square is 529; 8, of which the square is 64; 7, of which the square is 49. Total of the four squares: 13,411.

M. Lebesgue, author of *Introduction à la Théorie des Nombres,* has admitted that he would have needed a fortnight to arrive at a similar result.

Inaudi indeed calculated at an astonishing speed. Thus in 1924 Maurice d'Ocagne had the notion of organizing a competition at the Société des Ingénieurs Civils to match the calculator with calculating machines of the period. Inaudi beat the machine in addition, subtraction, raising to powers, extraction of roots and in most of the multiplications. It was in multiplications of five figures upward that the machine showed itself quicker than the man.[1]

In the other cases Inaudi had already given the answer to the problem before the machine had finished taking down the factors. Moreover, like the majority of lightning calculators, Inaudi gave the day corresponding to any date almost instantaneously, which the machine could not do.

Alfred Binet, who studied Inaudi from a psychological point of view, has shown that the calculator was essentially an "auditory" type and that his memory was very specialized. While he had the faculty of remembering hundreds of figures he was incapable of repeating more than five or six letters enunciated in a certain order: a, r, g, f, s, m, t, u, for example. He displayed the same inability to recite two lines of verse or prose. On the other hand he could maintain a conversation and answer questions wittily and to the point, while solving a complicated problem in his head.

[1] Electronic "brains" and modern calculating machines make it possible to carry out, at the speed of light and in infallible fashion, the most complex calculations which would otherwise occupy teams of specialists for weeks or even whole years. These machines can give the product of two numbers of ten figures in less than 50 millionths of a second.

These have sometimes been described as "thinking machines." In reality this is true only within the limits set to the invention by man and in relation to the final result which he has assigned to it. It is certainly possible to make these machines carry out reasoning processes, such as syllogisms, on condition, of course, that the appropriate symbols are chosen and the right circuits established. But between this and genuine "thinking" there is a gulf which it is impossible to bridge.

Pericles Diamandi

The lightning calculator, Pericles Diamandi, who was a contemporary of Inaudi, does not belong to the group of calculators of less than average intelligence or of slight education. On the contrary he was a highly cultivated man, but as he was studied at length by Alfred Binet and sought to match himself against Inaudi we will mention him here.

Diamandi, born in 1868 at Pylaros in the Ionian Islands, came of a family of grain merchants. During his student period he was consistently first in mathematics. In 1884 he left school and began work in the grain business. It was at this time that he realized that he had exceptional powers of mental calculation. He cultivated these and discovered methods of simplification. At the same time he learned Rumanian, French, German, English and two other secondary languages, and wrote verses and novels. Having read by chance in a newspaper one day an account of a performance by Inaudi, he was seized with a desire to emulate him and gave performances himself of mental calculation in Greece and Rumania. He came to Paris in 1893 in order to match himself with Inaudi; but for various reasons the encounter did not take place. Diamandi soon had himself presented to the Académie des Sciences, before whom he wished to demonstrate his aptitudes as a calculator. The Académie entrusted the examination of the young man to the same committee which had tested Inaudi. An investigation was made, and at its conclusion Professor Charcot published a study of Diamandi in the *Revue Philosophique.*

Later Alfred Binet revised and completed Charcot's work and also published the result of his own investigations in the same review. These showed that Diamandi had approximately the same capacities as Inaudi but that his memory was "visual" whereas that of Inaudi was, as we have seen, "auditory." Diamandi "saw" the figures in his head, in front of the frontal lobes, as if they had been written up on a screen. They seemed to remain fixed there until by an effort of will he caused them to be effaced.

When, instead of seeing figures, he heard them, he was obliged to operate a sort of transformation from an auditory to a visual image before starting to calculate. Diamandi therefore asked that for preference the numbers on which tests were to be based should be written down on a piece of paper or on a blackboard. For example he asked one of the assistants to write any five numbers of five figures on a blackboard:

4 9 3 5 7
8 0 2 4 6
9 5 3 1 4
2 7 6 9 5
7 6 2 3 2

He looked at them for a few seconds, then, from memory, repeated them at will from bottom to top, from top to bottom, from left to right, from right to left, diagonally and in any other desired direction. Spectators had the impression that these figures had been photographed on his memory and that he read them within himself with the same facility as if he were reading them from the blackboard.

Like Inaudi, Diamandi could perform arithmetical operations very rapidly. He was asked, for example, to multiply a number of fifteen figures by a number of four figures. He dictated the answer at the end of half a minute. He took about two minutes to extract the square root of a number of ten figures. At the end of the proceedings he repeated from memory the figures concerned in all the operations he had been carrying out and which were inscribed on the blackboard, first in their correct order, then in the reverse order, without hesitation, without error, and so fast that it was almost impossible to follow. On several occasions he had to be asked to repeat them, by so much did his inward eye outdistance the normal eye of the examiner.

Diamandi could also give instantaneously the day of the week for any given date, but whereas Inaudi performed a calculation to obtain this result, Diamandi had devised a table on which the dates forming the last hundred years were inscribed in circular form. On this table a sort of grille turned, carrying the names of the months and days of the week. To know the day of a given date he had only, by turning the grille, to bring the name of the month together with that of the year in question. The name of the day appeared automatically. But Diamandi had no need to have this apparatus before his eyes. He had engraved it on his memory, and upon that mental screen he could read as upon a real table.

Rightly or wrongly, Diamandi believed that the faces of individuals represent one of the ways in which their psychology is expressed. Thanks to his exceptional visual memory he had formed a mental collection of the faces of innumerable people whose characters he knew; thus when he found himself confronted by people unknown to him he could, he said, describe their state of mind.

In this form of exercise his successes appear to have been remarkable and many who consulted him declared themselves absolutely astounded by the disclosures he made to them about the most secret inclinations of their minds, but it is clear that in this field a check is almost impossible.

"During my ten years of travel throughout the world," he confided to Gaston Méry, director of the *Echo du Merveilleux,* "I had opportunities of observing faces of all categories. In this way a sort of cinematographic museum was formed in my mind, extending over an immense variety of types. When I study a face its image evokes all the corresponding images, and these stand out as from a group and appear before me; I have only to compare. I know that this wrinkle, that fold of the lips, that expression of the eyes, that shape of nose are the mark of this or that psychological disposition, the stigmata of this or that way of thinking or feeling. It is true that I had to work by trial and error for a long time, and that for certain traits of character I still do. But on balance, if in face reading I have not succeeded in attaining the absolute precision of a mathematical operation, I have achieved a relative precision, which I shall try to improve more and more by new observations, but which already seems to me to be far from negligible."

One of his sisters, Uranie, was also a lightning calculator. It was at the age of seven that she noticed her special aptitude. Her brother's first successes later encouraged her to cultivate this gift. Her visualizing power extended not only to figures, which appeared to her as colored; similarly she attributed colors to the letters of the alphabet and the names of the days.

Here are the colors she associated with figures, as well as certain letters and days of the week:

0 White (like the letter O)
1 Black (like the letter I)
2 Bright yellow (like the letter S and the word "Sunday")
3 Vermilion (like "Wednesday")
4 Very dark brown
5 Royal blue (like "Saturday")

6 Bright yellow (as for 2, but darker)
7 Very dark navy-blue
8 Gray blue
9 Sepia (like F and "Thursday")

Uranie Diamandi affirmed that she remembered most easily numbers containing figures with light and conspicuous colors framed by figures of dark or dull colors. She found that in these cases the association of colors and figures aided her memory.

"For instance," she said, "104 (black, white, brown) is easy to learn and to remember because 0, which is white, is here placed between two dark colors. In the same way 129 (black, bright yellow and sepia) is better learned and remembered because of the contrast."

The mental arithmetic carried out by Uranie Diamandi was somewhat similar to that performed by her brother. Five rows of five figures were inscribed on a blackboard. She looked at the square for about a minute then, turning her back, recited it in all directions and named any figure whose position in the square was given her. She added up the rows mentally, carried out the most varied operations upon the numbers: subtractions, multiplications, divisions, raising to powers, etc., and was moreover able to raise a figure up to the twentieth power, which corresponds to a number formed from between 7 and 20 significant figures.

Louis Fleury

Inaudi, who has been justly considered for fifty years as the giant of lightning calculators, has had rivals worthy of him in recent years: Louis Fleury, Mlle. Osaka and Maurice Dagbert.

Louis Fleury, born near Belfort on April 21, 1893, was afflicted from birth with a double ophthalmia which made him completely blind. Abandoned by his parents at the age of eighteen months, he was placed by the Public Assistance in a family of small farmers. At the age of ten he could barely walk and could neither wash nor dress himself. An effort was made to give him some education at the school for the blind at Arras; it appeared that calculation was his weak point: he learned addition and subtraction with difficulty and multiplication with even more difficulty, while the mechanism of division remained totally incomprehensible to him. In contrast to most of the great calculators, Fleury was thus a mentally retarded case where elementary arithmetic was concerned.

At fifteen, considered as ineducable, he was placed in a home for incurables.

"He had been there two months," writes Dr. Osty, "when he received a sudden and violent shock. A man of about forty, his neighbor at table, uttered a loud cry and rolled on the floor in an epileptic fit. In his night of blindness, the convulsions of the patient and the cries of those about him assumed terrifying proportions for Fleury. The emotional shock was so great that he became ill for several days. This first and violent terror long haunted his mind, like an agonizing obsession.

"A mental transformation resulted. For psychologists this is perhaps the most interesting aspect of Fleury's case.

"Seeking within himself for a cure for his obsession, he had the idea of concentrating upon work which was the most absorbing for him because it was the most difficult. He set himself to do mental additions, subtractions and multiplications which up to then he had been able to do in the writing of the blind only up to a certain degree of complexity. It was miraculous! All the calculations he attempted resolved themselves with wonderful ease, rapidity and sureness. Even division, that irreducible fortress, was as easy as the other operations.

"From then on the abstract world of figures became his real inward life; his mind worked therein without effort and with delight. Mental calculation became his great distraction, a kind of sport, the intellectual sport of a man whom circumstances and blindness had condemned to live most of his time sitting down. A sport however without much real progress, for all that he undertook he achieved. His practice in calculation was not so much a forward march towards greater accuracy or facility, as the exploration of the extent of his capacities.

"And this gift for calculation, which had emerged from the depths of a psychotic crisis, brought about a general improvement. His mind, hitherto clouded, became wholly cleared. This was manifested by a feeling of greater aptitude for learning and a desire to educate himself."

Fleury in fact asked to return to a blind school, but the Public Assistance would not agree to this. He then resolved to escape at all costs from the depressing futility of the environment in which he lived, and to this end he simulated madness. He was admitted to an Armentières mental hospital, where it was rapidly perceived that he was not mad, but that on the contrary he possessed exceptional powers of mental calculation. In order to widen the scope of his capacities it was explained to him what the square of a number is, and he at once calculated the squares of numbers of three and four figures. Next, a square root was defined for him, but without indication of its method of extraction. Within a few days Fleury discovered a method enabling him to extract square roots for four figure numbers mentally and without mistake.

When he came of age, Fleury left the hospital at Armentières and gave some exhibitions in France; then he went to England and to the United States, where he gave performances in schools, theaters and travelling circuses. He returned to France in 1927, and it was at this time that he was examined at the International Psychical Institute by Dr. Osty and his colleagues. The program for the tests was the following: M. Sainte-Laguë, who held a mathematical degree and was a professor at the Conservatoire National des Arts et Metiers of Paris, played what amounted to the role of examiner. Before the demonstrations he prepared a certain number of operations to be carried out, as well as their answers. The questions were put orally at a rapid rate; as soon as one answer was supplied by Fleury, another problem was put. Dr. Osty measured the time for each calculation with a chronometer.

Here are some of the questions, with the answers and the duration of the mental calculation:

Multiply 553 by 88. Answer: 48,664, in 2 seconds.

Multiply 649 by 367. Answer: 238,183, in 10 seconds.

Divide 5,364 by 43. Answer: 124, remainder 32, in 4 seconds.

Divide 20,700 by 48. Answer: 431, remainder 12, in 3 seconds.

Raise 5,287 to its square. Answer: 27,952,369, in 10 seconds.

Raise 94 to the power of 4. Answer: 78,074,896, in 15 seconds.

Raise 2 to the power of 20. Answer: 1,048,576 in 20 seconds.

Raise 2 to the power of 30. Answer: 1,073,741,824, in 40 seconds.

Extract the square root of 13,250. Answer: 115, remainder 25, in 4 seconds.

Extract the square root of 222,796. Answer: 472, remainder 12, in 12 seconds.

Extract the cube root of 456,609. Answer: 77, remainder 76, in 13 seconds.

Extract the fifth root of 1,935,752,415. Answer: 72, remainder 834,783, in 3 minutes 10 seconds.

Problems of a different kind were also put to him. Thus M. Sainte-Laguë supplied the total of a number raised to its cube added to another number of four figures. Fleury had to give the number which had been cubed and the number which had been added to this cube.

The number given was 707,353,209. Answer: 891 cubed and 5,238, in 28 seconds.

The number given was 211,717,440. Answer: 596 cubed and 8,704, in 25 seconds.

Another problem: break down 6,137 into four numbers which are perfect squares. Fleury gave three answers successively.

First answer: 5,476, square of 74; 400, square of 20; 225, square of 15; 36, square of 6, in 2 minutes 10 seconds.

Second answer: 6,084, square of 78; 36, square of 6; 16, square of 4; 1, square of 1, in 10 seconds.

Third answer: 5,776, square of 76; 225, square of 15; 100, square of 10; 36, square of 6, in 1 minute 20 seconds.

Finally, for any date, past or future, by the Gregorian or the Julian calendar, Fleury gave the day of the week almost instantaneously.

While the majority of lightning calculators are "visuals," Fleury is of the "tactile" type, which is very rare. He said that he "felt the outlines of imaginary cubarithms passing beneath his fingers," that is to say the embossed counting symbols used by blind people. "When he was carrying out an operation," wrote Dr. Desruelles, who studied Fleury at the asylum at Armentières, "his fingers moved with extreme rapidity. With the right hand he grasped the fingers of the left hand one after another; one represented hundreds, another tens, a third units. He moved his fingers feverishly over the lapel of his jacket and it was curious to watch him using these tactile images to obtain sensations corresponding to those he would have had in touching cubarithms."

Like all lightning calculators, Fleury employed some simplifying processes, but, to be frank, these were in no way remarkable and in consequence do not deserve mention. On the other hand it is of interest to note the manner in which he proceeded to extract square or other roots, to break down a number into perfect squares and to give the day of the week corresponding to a given date. But let us emphasize at once (and this observation applies to a great number of operations performed by lightning calculators) that these processes, which are carried out practically instantaneously, are for the most part unconscious. It is thanks to analysis that it has been possible to reconstruct them. Take for example the extraction of the square root of 1,526. Fleury's technique, arrived at after successive trials, is the following: the root required is greater than 30 and smaller than 40; 30 squared gives 900—this number is too small; 33 squared gives 1,089; 37 squared is equal to 1,369; these two squares are both lower than 1,526, and their roots will not fit; 40 squared gives 1,600; this number is too high. The square root is probably 39. In fact 39 squared is equal to 1,521. The correct answer is therefore 39, with a remainder of 5.

The breakdown of a number into four perfect squares is an application of the preceding process. Take, for example, the breakdown of 12,315 into four perfect squares. Fleury seeks first for a perfect square fairly close to 12,315. This is the case with 10,000, square of 100, but 10,000 is not sufficiently close 12,315; 110 squared gives 12,100 and this number can serve; thus there remains 215 to be broken down into three squares. Fleury now seeks a number which, squared, gives a result near to 215, but he finds that the remainder cannot furnish two perfect squares; he therefore leaves it and makes a second attempt. He tries 105 squared, which gives 11,025; there remains 1,290, which must be split into three squares; he next takes 35, which he squares, giving 1,225; the remainder is 65; he then sees easily that the square of 7, which is 49, and the square of 4, which is 16, together total 65. The four numbers sought are therefore: 11,025, 1,225, 49, 16.

Where a day of the week to correspond with a date is needed, this, according to Dr. Osty, is Fleury's method. It is necessary to remember:

(a) that for a date before 1582 (Julian Calendar) the 1st of January 1582 was a Monday;
(b) when a date later than 1582 (Gregorian Calendar) is involved, what day of the week the 1st of

January of the present year is (example: January 1, 1927 was a Saturday);

(c) that the first day of the eleven other months of the year correspond to this or that day of the first fortnight of January;

(d) that the following correctives must be used: 1. forward displacement of one day for the months of leap years, less February; 2. in consideration of the displacement of one day in the leap years, to make a reduction for the Gregorian Calendar of 12 days in every twenty years or of 61 days in every hundred years, since a date must fall on the same day every seven years; 3. for the Julian Calendar reduce by 12 days for every twenty years and by 62 days for every hundred years.

It should be noted that Fleury worked out all these ratios in his head while at the Armentières asylum. It can easily be imagined what their discovery represented in terms of research and effort and in innumerable comparisons.

Two examples given by Dr. Osty show how Fleury used these points of reference.

Question: What was the day of the week of August 13, 1911?

Base of reference: January 1, 1927 is a *Saturday.*

1st stage: Determine the day of January 1, 1911. Given the 16 years of interval between 1927 and 1911, refer back twenty years, that is to 1907, and reduce by 12 days to find January 1, 1907, which gives a *Tuesday.* And, as each new year involves a forward displacement of one day (except for leap years: 2 days) it follows that January 1, 1908 is a Wednesday; January 1, 1909 a Friday (2 days for 1908, a leap year); January 1, 1910, a Saturday; January 1, 1911 a *Sunday.*

2nd stage: To know which day August 1, 1911 was. August 1 of a year falls on the same day as January 3 when the year is not a leap year, and January 4 when it is a leap year. Therefore, as January 1, 1911 is a Sunday, January 3, 1911 is a Tuesday and August 1, 1911 is also a *Tuesday.*

3rd stage: Determine the day of August 13, 1911: August 1, 1911, Tuesday; August 15, 1911, Tuesday; August 13, 1911, *Sunday.*

Now, and this is what is worthy of attention, all these points of reference and all these reasoning processes passed through Fleury's mind *in less than one second,* since he gave the correct answer at the end of that time.

Another question: What was the day of the week of September 19, 139?

Basis of reference: January 1, 1582 is a *Monday.*

1st stage: Apart from the points of reference already mentioned, Fleury knew, since he had worked it out, that, in the Julian Calendar, a date falls on the same day of the week every 28 years. This enabled him to go straight back 1,400 years (1,400 being divisible by 28). Hence 1582—1400 = the year 182. January 1, 182: *Monday.*

2nd stage: Find the day of the week of January 1, 180, distant from the year 140 by twice 20 years. January 1, 182, Monday; January 1, 181, Sunday; January 1, 180, *Friday* (reduction of two days for leap year).

3rd stage: Find the day of the week for January 1, 140. From 180 to 140 there are twice 20 years, hence a reduction of twice 12 days, which gives January 1, 140: *Thursday.*

4th stage: Find the day of the week for January 1, 139. This day will be one day before that of January 1, 140, therefore: *Wednesday.*

5th stage: Find the day of the week for September 1, 139. September 1 of a year always falls on the same day of the week as January 6, except in leap years when it falls on the same day as the 7th. Therefore September 1 is a *Monday.*

6th stage: September 1, Monday; September 15, Monday; September 17, Wednesday; September 19, *Friday.*

Here, the answer was given after *4 seconds.*

Mlle. Osaka

If Fleury was, as we have said, of the "tactile" type, the woman lightning calculator Mlle. Osaka belongs, like most of the great calculators, to the category of the "visuals."

This girl, whose borrowed Asiatic name in fact disguises a Frenchwoman born a few miles from Bagnères, was a mentally retarded child. She began walking and talking only at the age of four and a half. Because she was sickly she scarcely attended school, so that at twenty-six she could only just read and write. Her knowledge of arithmetic was confined to addition. One day she was present at a performance given, if not by a lightning calculator, at least by a calculating virtuoso, and without knowing why, she felt that she would easily be able to perform the same feats.

"Let us reflect," Dr. Osty justly notes in this connection, "upon the important psychological significance of this fact, occurring to so many people who have been ignorant of their own exceptional gifts up to the day when an accidental circumstance has arisen to stir in them the sense of their special capacities in this or that direction: artistic, literary, scientific or others. In such cases it is as though the subconscious mind is made aware of the individual's hidden resources, and under the impulse of circumstance, succeeds in forcing its way across the threshold of consciousness and imposing upon it the vital spark."

However this may be, Mlle. Osaka, impelled by this strange certainty, set herself to learn the rules of calculation of which she was ignorant, that is to say subtraction, multiplication and division, but found herself held up by this last operation, of which she was never able to understand the workings. Two facts of great importance however soon made themselves evident to her mind. She found on the one hand that she could calculate with extreme rapidity, and on the other that she could retain in her memory the numbers she had manipulated in her head. This second discovery induced her to turn her training in another direction. She stopped trying to study calculation itself and sought instead to memorize larger and larger numbers. From this moment on, her progress was extremely rapid, so that she could fulfil her secret desire to exhibit herself in public. She perfected her aptitudes, learned by heart a colossal mass of numbers which she calculated on paper: the powers of numbers of 1 and 2 figures, up to the 10th; the powers of numbers of 3 figures up to the 7th or 8th, the numbers of hours, minutes, seconds corresponding to ages, etc.

In these conditions her mental numerical luggage was literally indestructible, and Mlle. Osaka was able to give the answer immediately and without a mistake, within the framework of her knowledge, to every question on powers or roots; and it was equally possible for her to give the number of seconds lived by a person of this or that age, etc., with the same facility. When she wishes to remember numbers she sees them as if they were outside herself; when a hundred or so figures are dictated to her it seems to her as though they were written up on a blackboard and are, as she says, "more legible than the real ones." Thus, if she is asked for the sixth power of 97, she will "see" all the multiplications which she has used to calculate the powers of 97, from the second to the sixth. If a board is covered with figures and she is asked what number is inscribed on the fifth line, Mlle. Osaka, who has only heard the figures called out, sees all the numbers on the board instantly and clearly. Her capacity for mental retention is so great that she can repeat them either in the normal order or backwards, with the greatest of ease. The following experiments carried out by Dr. Osty at the International Psychical Institute give an idea of her extraordinary mnemonic capacities.

Dr. Osty asked for the square of 97, then for the 10th power of the same number, both of which the calculator gave instantly. After that he asked for the sixth root of 402,420,747,482,776,576, then for the square root of the same number and again these were given at once and correctly. This done, he wrote down, absolutely at random, a succession of a hundred figures and then read them out at a speed of approximately one figure per second.

"When I had finished this test," writes Dr. Osty, "Mlle. Osaka repeated the hundred figures in the order in which they had been spoken. About forty-five minutes afterwards, and when we had been talking about a number of things, I said to Mlle. Osaka, who had no reason to expect the question:

" 'Could you repeat to me the hundred figures which I dictated to you nearly an hour ago?'

" 'Very easily,' she replied.

" 'Would it be possible to do this backwards?'

" 'I will try.'

"And she succeeded."

In the course of a demonstration conducted with a small audience, twenty people each wrote a number on a piece of paper, each paper bearing a control number from 1 to 20. The papers were mixed up, drawn at random, and the numbers on them read aloud, then written as they were called on a blackboard containing twenty sections numbered from 1 to 20. They varied from millions to nonillions.

Mlle. Osaka, who was standing facing the public, heard them but did not see them. She turned her back to the board which, in fact, could have been removed.

In order to complicate the test Mlle. Osaka asked that a few problems should be put to her before she repeated the numbers on the board. One of the audience asked her for the square and then for the 10th power of 27, another for the 10th power of 55, then to go through all its powers in descending order. It was child's play for this girl to answer these questions.

Then Mlle. Osaka was asked to repeat the twenty numbers written on the blackboard in the order from 1 to 20, which she had not heard since the order of inscription had been determined by chance. This she did immediately with extraordinary rapidity and without any mistake.

"This was followed," writes Dr. Osty, "by requests for the powers of 3 figure numbers. Someone asked for the 2nd, the 3rd, the 4th, then the 5th powers of 221. These were given exactly and without delay.

"To those who had supplied the date of their birth, Mlle. Osaka at once gave the number of days, hours, minutes and seconds they had lived, leap years being taken into account. One person proposed the mental multiplication of 624,987 by 2,358. Mlle. Osaka did this at slow speed, figure by figure. Forty-eight figures were thus produced in seven minutes without apparent effort and without error. Mlle. Osaka was then asked if she could repeat the numbers which had been inscribed on the blackboard, starting at the end. She did this immediately, figure by figure, from section 20 to section 5, then by groups of three figures for the remaining sections. She was then asked to state again the numbers of section 6, section 13, etc. Another person asked her to repeat the number in section 7 backwards. The correct answers were immediately given them. At this moment Dr. Moutier, who had earlier asked for the multiplication of 624,987 by 2,358, asked Mlle. Osaka if she could again give the result of this multiplication. The 48 figures were at once enumerated at very high speed. M. Morice, an architect, asked, as an addition to the blackboard test, that this should be turned face to the wall and that Mlle. Osaka should again repeat the numbers. To amuse herself, the girl repeated them all as fast as she could utter them."

These feats appear really prodigious when it is remembered that their execution is based upon a colossal memory for numbers. To remember thousands of numbers, each formed from 15, 20, 30 and up to 40 figures, to bring them out instantly and exactly from the depths of the subconscious, is a terrifying operation, which as I believe, and shall emphasize later, borders on the paranormal.

And here is another strange aspect of the extraordinary memory of Mlle. Osaka, an aspect which we have already encountered with other lightning calculators. Apart from numbers, she has difficulty in remembering current facts and things she learned at school. She has never been able to learn the correct order of the letters of the alphabet.

Maurice Dagbert

M. Maurice Dagbert, the lightning calculator who came into prominence at the conjurors' congress held in Paris in 1947, and who later displayed his full capacities at the Lausanne Congress in 1948, has certainly not the gigantic memory of Mlle. Osaka; nevertheless his powers of memory are exceptional. Moreover his capacities as a mental calculator are such that they seem at least to equal Inaudi's. Among other feats performed before the Académie des Sciences, he extracted a fifth root (answer: 243) in 14 seconds; a seventh root (answer: 125) in 15 seconds; a cube root (answer: 78,517) in 2 minutes 15 seconds; a fifth root (answer: 2,189) in 2 minutes 3 seconds; and raised 827 to its cube in 55 seconds.

Here is the report on M. Dagbert by MM. Gaston Fayet, Jean Chazy and Joseph Pérès, taken from volume 220 of the *Comptes rendus hebdomadaires des séances de l' Académie des Sciences:*

"At the request of the permanent secretaries we have examined the 'mental calculator' Maurice Dagbert, who wishes to be presented to the Académie. The results of this examination have been conclusive and appear to us to be worth publishing in the *Comptes rendus.*

"M. Dagbert's powers of calculation seem to be comparable with those of Jacques Inaudi, presented to the Académie by Darboux in 1892. Like Inaudi, M. Dagbert is served by an exceptional memory. He has informed us that he works out figures with the aid of extremely vivid images which he obtains by shutting his eyes or by staring at a white object (the ceiling of the hall in which he is operating, for example). He sees the figures appear, as they are announced to him, as though he had written them himself on a blackboard.[1]

"In the course of the examination, which lasted 2 hours 30 minutes, M. Dagbert had occasion to carry out varied calculations (perpetual calendar for Gregorian or Julian dates, multiplications, powers and extraction of roots). The details have been recorded in a report on the meeting which will be preserved in the Académie archives.

"M. Dagbert has had only a primary education and his knowledge of mathematics and elementary algebra are approximately nil. His taste for calculation was very precocious and a visit he made to Inaudi when he was 14 led him to make a personal effort which evidently has been very fruitful. He has told us that he has discovered for himself the rules he uses in his calculation, rules which for him are purely empirical and of which he does not explain the reasoning. He has given us the example, a particularly simple one, of the rule he employs to evaluate the cube of a number of two figures: he utilizes two key numbers, determined by the figure of the units u, which he knows by heart but whose origin eludes him. It is recognized at once that the first key number x is the figure of the units, the second y the number of the tens in $3u^2$. Dagbert's rule thus appears as the result of the development of the binomial $(10d + u)^3$, the calculation being directed to obtain the various figures of the result successively: u^3 gives the figure of the units and the remainders to be carried; xd, to which the amount carried forward is added, gives the figure of the tens and fresh remainders; $(3u + dy)\,d$, by adding the remainders, the figure of the hundreds; finally, the carrying forward accomplished, d^3 is added, giving the thousands of the answer.

"It is not surprising in these conditions that M. Dagbert, whose speed and powers of calculation are remarkable in the fields he knows so well, should be put off the track by quite simple questions (like those put to Inaudi at the time of his presentation in 1892) but which necessitate some algebraical transformations.

"After the demonstration was over, M. Dagbert was presented to the members of the Académie and carried

[1] M. Dagbert can, with absolute accuracy, represent the figures to himself as if they were written in white on a blackboard. His mental pictures are less clear if the figures are red on a blue background, and even less clear with yellow figures on a green background.

out before them some of the very complicated mental calculations of which he is capable."

In his exercises in public the arithmetical operations he performs mentally overlap, so that the cascades of figures which pour out over the audience, almost without a break, form a strange medley. First one person is invited to give his age, then five numbers of two figures are proposed by the public. A little while afterwards the calculator gives the third power of the first number, the fourth power of the second number and the fifth power of the third; he then stops to indicate to a spectator that he has lived for so many hours, minutes and seconds, and shows, by a calculation on the blackboard, that he has taken leap years into account. He rounds off by supplying the sixth and seventh powers of the last numbers, these two answers, be it noted, having 11 and 13 figures respectively.

More difficult feats are then proposed: raising 3 figure numbers to their cubes, then the extraction of roots. Someone for instance gives a 15 figure number, another one of 19 figures and finally dates on either the Gregorian or the Julian calendar; all these are given simultaneously. Instantly the artist indicates the day of the week to which these correspond, then announces the cube root of the first number and, partially, the fifth root of the second. He replies to further requests for dates. Finally he gives the complete cube root of the second number. Similar operations follow with the greatest rapidity, interspersed by replies concerning the dates of Easter, Ascension, Whitsun and the phases of the moon.

Finally he ends the performance by repeating all the numbers that have been announced during the session, that is to say, some hundred and fifty figures.

It should be added that in the course of his exhibitions M. Dagbert often plays brilliant pieces on the violin while he is solving complicated calculations in his head. Thus while he has been playing a fantasy from *Il Trovatore* with remarkable skill I have seen him carry out the extraction of twenty cubic roots of 3 figures and the multiplication of a 5 figure number by another number of 5 figures. The total operation lasted seven minutes. As he put down his instrument the calculator gave the twenty-one solutions straight off, absolutely without error. At no time did he make use of pencil or paper, even to note the terms of the problems.

The Psychology of Lightning Calculators

After this brief study of the principal lightning calculators, let us consider in what conditions their aptitude makes its appearance and by what mechanism it is developed.

First of all it seems that the gift of calculation is not inherited. The only two exceptions known are those of Bidder and Diamandi. The first passed on his gifts to his children and grandchildren, while Diamandi had one brother and one sister who possessed aptitudes for mental calculation similar to his.

Among all of them the gift has appeared spontaneously, without any outside stimulus. Indeed, many of the lightning calculators were born of poor and even very poor parents who paid little attention to the teaching or education of their children. It should be added, as we have already pointed out, that several calculators were at first considered to be backward children. At the age of 17, the Belgian calculator Oscar Verhaeghe expressed himself like a baby of 2.[1]

It has been mentioned earlier that Zerah Colburn displayed a sign of degeneracy: an extra finger or toe to each limb. Another lightning calculator, whom we have not mentioned so far, was Prolongeau, who was born without arms or legs. Mondeux was an hysteric.

To sum up, neither the external environment in which the lightning calculators developed, nor their general intelligence, at least in the case of most of them, gives us the explanation of their faculty: they have had neither the influence of a teacher nor an example; they have not been drawn into their way of life by guidance or by a normal education; their level of intelligence has often been very much inferior to their extraordinary arithmetical capacities. As Alfred Binet has rightly said: "In the birth of their faculty, there is something which resembles a sort of spontaneous generation."

[1] Oscar Verhaeghe, born on April 16, 1926 at Bousval (Belgium), in a family of modest civil servants, belongs to the group of calculators whose intelligence is well below the normal. The raising to various powers of numbers formed from the same figures is one of his specialties. Thus, 888,888,888,888,888 is raised to its square in 40 seconds and 9,999,999 to its fifth power in 60 seconds, the results involving 35 figures. Oscar Verhaeghe has been put through a certain number of tests by various learned groups and by the eminent mathematician Kraichit, of Brussels University.

Another trait which characterizes the lightning calculator is precocity. In the cases of Gauss and Whateley the faculty made itself manifest at the age of three years. It is reported that Gauss's father was in the habit of paying his workmen at the end of the week and that he added to their wages a bonus for any extra hours they had done, based on the total earned for each day. One day, when Gauss senior had just finished his calculations and was preparing the money, the child, who was then barely three and who had been following his father's operations without being noticed, cried out:

"Father! Father! You've done it wrong. This is the amount." The sum was done again carefully and to everybody's astonishment it was discovered that the right amount was indeed that given by the little boy. In the same fashion Ampère brought off quite lengthy mental operations at the age of four, when he knew neither letters nor figures. He used nothing but a few pebbles or beans. A story told by Arago is enough to demonstrate the point to which love of calculation had taken possession of the child. Maternal tenderness having deprived the young Ampère of his beloved beans during an illness, he replaced them with fragments of a biscuit which he had been given after a three days' fast.

Ampère, as we know, became one of the greatest French physicists, but, strange to say, in proportion as his knowledge of classical mathematics and science developed, so he progressively lost his aptitude for mental calculation.

It was the same with Whateley, who notes: "My ability to calculate displayed one peculiarity. It showed itself at the age of four and lasted three years. I could do the most complicated additions in my head, and a good deal more rapidly than others who did them on paper, and no one ever found any mistakes in them. But by the time I began going to school, my ability to calculate had vanished and since then I have always been very weak in mathematics."

Safford, according to M. Scripture, could do multiplications in his head with answers running up to 36 figures at the age of five. As he had remarkable aptitude for classical mathematics, he became a professor of astronomy, but entirely lost his faculty for lightning calculation.

On the other hand the Swiss, Leonard Euler, who began calculating at the age of five and who, according to his biographer Lacroix, "deserves to occupy in mathematics the place which Voltaire holds in literature," retained his prodigious facility for mental calculation until extreme old age. Possessed of an encyclopaedic mind and gifted with a colossal memory, he not only solved the most complex problems of analysis or geometry in his head, but also knew by heart the Aeneid of Virgil, his favorite author, and had a profound knowledge of physics. chemistry, zoology, botany, geology and medicine as well. He was also well versed in history and the Greek and Latin languages.

In 1937, an English Sunday newspaper published a Reuter report from Vienna that Meho Focie, aged 5, the son of a bootmaker, had amazed mathematics teachers in Zagreb. He had not learned to read or write, but he could solve any complicated problem in his head. He was able to multiply or divide six figure numbers and find their square roots without a mistake in a few seconds. A mathematician, watching him at play, asked him: "I am fifty-one today; how many days have passed since my birth?" Without stopping his game the child gave the exact answer, taking leap years into account.

Colburn, Prolongeau and Inaudi began calculating at the age of six, and Bidder, Mondeux and Mangiamele, the less precocious examples of the series, at the age of eleven. Alfred Binet has suggested eight as the average age for the first manifestations of the gift of lightning calculation. However that may be, what should be especially noted on this subject of precocity is the number of them who began to calculate *before knowing how to write figures*. Thus they worked out numbers in their heads *without understanding the meaning of a written figure*. This is a rather mysterious fact, difficult for an adult accustomed to calculate with these precise and convenient symbols to understand.

An exceptional memory for numbers, the ability to keep them stored in the consciousness in a manner bordering on the supernatural also characterizes the lightning calculators. It is in both these characteristics specifically that they surpass ordinary mortals, and in the second that they resemble hypnotic or psychic subjects in whom sensation can take the form of hallucination.

The memory of the lightning calculator is generally a visual one, but it may also be auditory, tactile or motory.

Mondeux and Colburn saw figures form before their eyes, as if traced by an invisible pen. Diamandi perceived them before his frontal lobes. Mlle. Osaka saw them, when they were dictated to her, as if written "in white upon a blackboard." It was the same for Dagbert. Dismer, the shepherd who lived not far from Stuttgart, and Pierre Annich, the herdsman from the Tyrol, who were also lightning calculators, were also "visuals."

Inaudi, on the other hand, was primarily an "auditory." "I hear a voice which calculates," he often said.

This voice, however, did not prevent him from following a conversation, or from simultaneously carrying out easier calculations than the main problem put to him, or from playing the flute. The mysterious voice continued its soliloquy and at the end of a certain time supplied Inaudi with the answer to an extraordinarily complicated calculation.

Fleury, as we have said earlier, was a "tactile."

Finally it seems that the majority of the calculators have also brought their "motor" memories into play, either by involuntarily executing certain movements with their hands or by silently articulating the figures. It was thus that Jean Hutzinger, a lightning calculator who had his hour of fame, constantly moved his lips while calculating. To sum up, it is very probable that a particular form of memory predominates in this or that case, but that all the memory forms play their part to a greater or lesser degree.

The Hidden Powers of the Mind

To explain the gift of mental calculation the classical psychologists invoke certain quite ordinary psychic qualities carried to a high degree. Their view is clearly expressed by the Swedish neurologist Dr. Jakobson in *Acta Medica Scandinava.*

"The lightning calculators," he says in substance, "are subjects gifted with a highly developed visual memory and an excellent memory of association, who are apparently unaffected by mental fatigue and are capable of concentrating their attention upon complex operations very rapidly and for a prolonged time. Impelled from a very early age by the development of a spontaneous faculty for mental calculation, they acquire progressively an automatic memory for arithmetical operations which is independent of all mathematical training properly so called. In this way they can greatly extend, by simple memory, the numbers of the multiplication tables of which they know the products by heart. Such subjects have no more need to reflect, that is to say engage even in a mental operation, in order to multiply two numbers of three or even four figures, than we have to multiply the ten first numbers in Pythagoras' table. Starting from this data they can easily reduce the most complex multiplications to a small number of operations of which they add up the products mentally. To do this they literally see the figures as 'written in the air.' Divisions are done in the same way, simplified by a method consisting in extracting from the dividend the squares, known by heart, of the round numbers into which the divisor can be resolved.

"Every arithmetical operation can be broken down in this way into a certain number of 'prefabricated' operations, of which the results are known by heart and which it is then only necessary to add up with the remainders. In essence, therefore, it is more a question of a very special form of memory, which has been developed by training, than of a genuine aptitude for calculation."

In reality this explanation is clearly inadequate. It explains neither their prococity nor their innate gift, nor the fact that it can show itself in mentally backward persons, nor the often prodigious character of the arithmetical memory of many of the calculators. Moreover it does not sufficiently emphasize the extent to which they perform their feats automatically.

When Inaudi hears a voice which calculates within him while he continues to converse or is consciously doing quite different calculations, when Dagbert plays a brilliant piece on the violin and during this time solves twenty-one complicated problems in his head, do we not see here appearing the "unknown guest" of Maeterlinck, emerging from the normal personality, superimposing itself upon it and proclaiming its independent existence? Certain calculators have moreover actually felt this "guest" living within them and have distinctly recognized it.

"From my childhood," writes Ferréol, "I calculated in an entirely intuitive manner, to such an extent that I frequently had the impression of having lived before. If I was set a difficult problem the answer sprang directly from my mind without my knowing at first how I had obtained it; I then looked for the method by starting from the result. This intuitive power of comprehension, which never made mistakes, developed in close parallel with the exercises which it was called upon to do. I still often have the impression that someone is beside me who whispers to me the answer I wish for, the paths I am seeking, and generally these are paths which no one, or almost no one, has trodden before me and which I should not yet have found if I had set out to look for them." (Reported by Moebius, *Über die Anlage zur Mathematik.*)

With this calculator the separation between conscious thought and the subconscious or psychic faculties seem to have been complete, the conscious control being only an accessory condition.

It was the same with Bidder:

"He possessed," writes M. V. Pole, "an almost miraculous faculty for finding, more or less intuitively, factors which when multiplied together gave this or that large number. Thus, given the number 17,861, he could see at once that this resulted from the multiplication of 337 by 53 . . . He said that he would not be able to explain how he did it; with him it was almost a natural instinct."

Much the same can be said of Verhaeghe, who is quite unable to explain the mechanics of the complicated mathematical operations he carries out with such extraordinary rapidity and accuracy. When he is asked how he sets about it, he replies: "I do not know. It comes to me like that."

Elsewhere, as with Buxton, Inaudi and Dagbert, there existed or exists an instant-by-instant collaboration between the conscious and the unconscious, a contact between the two levels of the mind, and it is certain that among calculators the artifices which enable operations to be simplified and which are apparently discovered by the conscious mind, are in fact, automatically introduced by the unconscious.

This is clearly implied in a statement by the "amateur" calculator M. Paul-Aumont Lidoreau, who uses a certain number of established tables in the extraction of cube roots.[1]

[1] M. Paul-Aumont Lidoreau was initiated into mental calculation from a very early age. When he was at school, he was already able to extract the roots of numbers of from 9 to 15 figures in his head by methods of his own which he later perfected.

Like all the virtuoso calculators he performs the perpetual calendar operation, and can calculate the age of a person in days, hours, minutes and seconds very quickly. But his specialty is the extraction of cube roots not ending in 5. In this class he has even beaten the record of Inaudi (which was 2 minutes 15 seconds for the extraction of a cube root of a number of 15 figures) by giving at the end of 20 seconds the cube root of a perfect cube of 15 figures and in 46 seconds that of a perfect cube of 18 figures.

"I do my calculations entirely in my head," he said, "without any cerebral fatigue. I am aided in this in an incomprehensible manner by my subconscious mind, and it is this, I suppose, which does the most important work. Thus, to extract the cube root of a 15 figure number, I have to perform an average of 12 to 15 operations in 20 seconds. Several of these calculations do themselves simultaneously in my mind without my knowing exactly how."

Added to these are other facts of a slightly different kind which demonstrate that in our psychic personality the unconscious is an active region, capable of creation. "It is not I who think," confided Lamartine, "it is my ideas which think for me." Similarly Alfred de Musset said: "You don't work, you listen; it is as if a stranger spoke in your ear." Schopenhauer has also defined the role of the unconscious: "My philosophic postulates have all produced themselves without my intervention, at moments when my will has been asleep and my mind not engaged . . . Thus my personal self has been a stranger to my work."

This subconscious activity also shows itself in the scientific field. "Instead of forcing myself to understand a proposition on the spot," Arago notes, "I admit provisionally that it is true; and the next day I am astonished to comprehend perfectly what seemed wholly obscure the night before."

Answering a question by Paul Valéry, Professor Langevin made a similar observation: "Each time," he said, "one thinks with intensity and has thus to some extent prepared the work of the subconscious, this latter continues of its own accord and sometimes gives warning when it has finished. I have very clear recollections of an inner shock informing me at a given moment that a question has been solved and that there is nothing left to do but to consciously express the result."

Joliot-Curie too stated that he had had "sudden illu-

In the course of a demonstration organized at the International Psychical Institute, M. Lidoreau extracted the cube roots of the following numbers in his head in a few seconds and without a mistake:

37,246,609
599,930,290,504
924,579,746,488
13,055,567,849,956,664

In contrast to the majority of lightning calculators, M. Lidoreau does not possess an exceptional memory for numbers.

minations" supplying him with the best means of producing and observing a phenomenon, with the immediate sensation that the method thus suggested was unique and that any other would be less simple. "This sensation," he says, "I can recall in at least two circumstances, of which one was the proof of the explosion of the uranium atom."

On his side Professor Jacques Hadamard reveals: "I have found a long-sought solution at the precise moment of a sudden awakening, caused by the motion of a car, and this solution has revealed itself to me in a direction quite different from that in which I had sought it until then."

Leaping over a mass of intermediate reasoning and outstripping years and even centuries of research, the unconscious is sometimes capable of bringing surprising truths to the notice of the conscious.

At the death of Pierre Fermat, one of the greatest mathematicians of the seventeenth century, there was found in his home a copy of the works of the Greek mathematician Diophantes, bearing in the margin the following annotation in Fermat's handwriting: "I have demonstrated . . . (he states a concept which it seems needless to quote here) but I cannot write down the demonstration, as the margin is inadequate."

"Now," notes Jacques Hadamard, "this demonstration for which the margin seemed to him too narrow, has been sought for in vain for three centuries past. Essential progress has been made along the lines of his theorem: it has already been demonstrated in limited cases. But for this it has been necessary to bring to bear a whole arsenal of algebraical theories scaffolded one upon another, and *of which none was known in Fermat's time,* of which none was even imagined at that time, and *no allusion by him indicates that he even suspected their existence.*"

Riemann, writing about his work on the distribution of prime numbers, sets out a series of properties and notes: "These properties are the consequence of a formula which I have not simplified sufficiently to publish." Mathematicians are still seeking for this formula.

Another example, as classic as the others and like them illustrating the creative power of the subconscious, is provided by certain discoveries of the mathematician Evariste Galois, who was killed in a duel at the age of twenty.

"During the night before the duel in which he was to lose his life," states Hadamard, "Galois wrote to a friend, summarizing a whole series of discoveries he had made. Among these was a mathematical statement which was quite clear to him, and is quite clear to us, but which could not have had any meaning in his time. As it stood it was not comprehensible. It became so only with the help of principles which were established a quarter of a century later.[1]

"Like Fermat and Riemann, Galois must have had a hidden knowledge of these fundamental principles, *inwardly perceived but not realized by his conscious mind.*"

Let us recall finally that it was to subconscious activity that Henri Poincaré owed the discovery of the fuchsian functions.[2] For some days he had been seeking to demonstrate that such functions could not exist when, "one evening," he states, "I drank some black coffee, contrary to my habits. I could not sleep; ideas surged up in clouds; I felt them colliding with each other until two among them came together to form a stable combination." The following morning the great mathematician wrote down the ideas which had arisen unbidden in his mind: the fuchsian functions were created!

From these various examples it appears that such unconscious acitvity has some analogy with the subconscious feats of lightning calculators. In the case of the mathematicians quoted above, as with the lightning calculators, the conscious sets out the conditions of a problem, and the result, which reasoning would not have been able to obtain rapidly or would never have obtained, springs suddenly from the subconscious.

What seems essentially to characterize the lightning calculator is that to a greater degree than ordinary mortals he can use faculties which are in some way innate and which probably exist in a latent state in every human being.

[1] The concept in question was submitted to the Académie des Sciences and judged to be completely incomprehensible by the eminent mathematician S. D. Poisson. Fifteen years after the death of Evariste Galois, the mathematician Liouville glimpsed the exceptional value of the proposition and published it in the *Journal de mathématiques.* "To understand the sense of it," he emphasized, "it is necessary to work at it for two or three months, doing nothing else."

Little by little mathematicians grasped its importance and demonstrated to what point it could play an important role in the coordination and clarification of principles of geometry, algebra and numerous branches of mathematical analysis.

[2] Henri Poincaré (1854-1912) developed earlier work on functions and differential equations by Immanuel Fuchs (German mathematician, 1832-1902). He used Fuchs' name to identify functional terms which remain invariable when substitutions are carried out in respect of variables with them in the same mathematical expression.

We have said advisedly "in every human being," for they have been found to show themselves in normal people, by the use of processes which separate the conscious from the unconscious artificially.

"A fair number of cases have been cited," writes Dr. Osty, "of people who, though unable to do the least mental calculation by ordinary means, have shown themselves to be lightning calculators when their subconscious has been placed in conditions where it can express itself directly, either by the use of so-called automatic writing or table-rapping, etc. I know a prominent Parisian who is exceptionally poor at any sort of calculation, even with pen and paper, yet who can supply immediate answers to any operation by unconscious writing; even if he has been rendered voluntarily inattentive to the problems, his ears have none the less registered unconsciously. I will wager that we would be surprised by the number and quality of lightning calculators we would discover if instead of demanding trivial and disconnected answers from tables and ouijas,[1] calculations were asked for instead. Any procedure which places our psychic potentials in conditions where they can express themselves could give birth to prodigies. Only one physiological perturbation, either imperceptible or evident, is needed to make possible on the functional plane of consciousness what is so often feasible where the unconscious alone is concerned, and a great calculator, precocious or the reverse, is thereby revealed."

Distinct from arithmetical operations proper but associated with the same order of intellectual exercises, were some curious experiments which were made at one time with a young medium of about nineteen at the Société Universelle d'Etudes Psychiques, then directed by Professor Charles Richet. These show quite clearly the astonishing powers of the hidden regions of the mind.

[1] The ouija is composed of a small ovel-shaped *planchette* with a pointer, on three short feet with or without casters, and of a horizontal board on which are set out the 26 *letters* of the alphabet, the 10 *figures,* and the words *yes* and *no.* The medium rests his hand on the planchette and his arm, which is alleged to be animated by a "spirit," pushes the apparatus over the board, in this way writing words and phrases, letter by letter. The phenomenon is most often a simple manifestation of psychic automatism. Indeed it can generally be put out of action by bandaging the medium's eyes and replacing the board he is accustomed to by a board on which the characters are arranged in a different order. When the subject has genuine qualities of clairvoyance the phenomenon can still succeed in these conditions, but this result is extremely rare.

It had been agreed to designate each letter of the alphabet by a figure: *a* by 1, *b* by 2, etc. A question was put to the medium, or more precisely to the personality which was supposed to be manifested through him: the reply via the ouija came in a few seconds: it was for instance 769. The ouija then gave the complete sentence; it was then seen that the sum of all the numbers representing the different letters of the sentence really did total 769. No doubt a little practice would help one to remember which number corresponded to which letter, in the order they appear in the alphabet, but the mental addition of all these numbers is certainly not an easy thing, especially when the calculation has to be made within only 2 or 3 seconds. Mnemonic exercises of a different kind were also carried out with lightning speed. One of the experimenters made a remark about the weather. At once the ouija traced out the following:

n e e g v a o r u o s l p e l d a s i a g p n z e.

As those present were not able to understand this type of logogriph, the personality explained, via the medium, that it was necessary to read first the 1st, the 3rd, the 5th letters and so on, omitting the letters of which the order number was indicated by an even figure; then on reaching the end of the letters to go backwards to the beginning, reading only the letters indicated by even figures. Thus one can read:

Ne vous plaignez pas de l'orage (Don't complain about the storm).

One of the onlookers remarked during the demonstration that the phenomena appeared to be authentic. The ouija immediately wrote an unintelligible series of letters which, when translated by the process described above, gave this sentence:

"I am pleased that you pay tribute to our good faith."

The relationship between these medium-produced logogriphs and mental arithmetical operations is easy to establish; in both there is a subconscious work of elaboration which appears almost impossible to accomplish, at any rate without a special cerebral disposition and long training. It will be noticed finally that sentences constituting a reply to an unexpected remark made by someone present are involved and that these could not have been prepared in advance.

The following experiment is recorded by Sir Oliver Lodge in his book: *The Survival of Man*. It, too, shows that subconscious intelligence can sometimes show itself superior to the normal intelligence of persons concerned. The account was written by one of Sir Oliver's former pupils, Miss C. M. Pole, later Mrs. Garrett Smith.

"In the early part of 1885," she writes, "I was staying at the house of Mrs. Q, and I and her daughter, Miss Q, B.A., London, used to amuse ourselves in writing with a planchette.[1]

"We had several planchettes but we could only get a response from one of them, which belonged to Miss Q. In the house with us were some eight or nine others, but for no other pair would the planchette act. The same one had formerly given good results with Miss Q. and another friend, but I had never written with a planchette before or since. We got all sorts of nonsense out of it, sometimes long doggerel rhymes with several verses. Sometimes we asked for prophecies, but I do not remember ever getting one which came true, and my impression is that generally when we asked for a prophecy the thing went off in a straight line —running off the table if we did not take our hands off . . .

"After some questions we asked it to write the equation to its own curve, in other words, to express mathematically the outline of the heart-shaped board. Planchette wrote something like this quite distinctly—

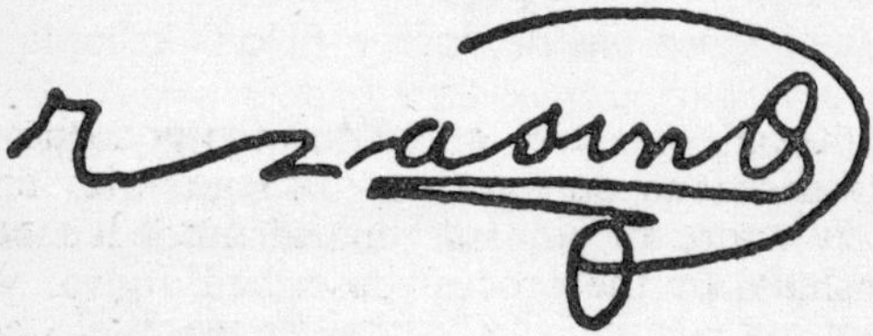

"The curl backwards always denoted that the answer was finished.

"We repeated the question several times, but each time the answer was the same, sometimes less distinct. We interpreted it as

$$\frac{r = a \sin \theta}{\theta}$$

[1] In this case the ouija was provided with a pencil.

"I knew just enough to be able to draw the curve represented by the equation. In my first try I made a mistake and believed the curve to be quite a different one, but afterwards I drew something like the following:

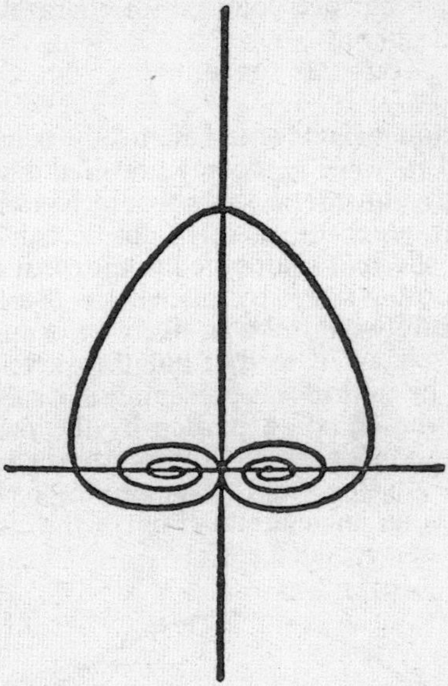

"We checked our result by taking the equation to the mathematical master at the boys' college, who drew the same sort of curve for us, but we did not tell him where we got the equation from.

"I cannot say whether the planchette we used was really exactly the shape of the outside curve; I should rather fancy that with the heart shape the resemblance ended. I am *quite sure* that I had never seen the curve before, and therefore the production of the equation could not have been an act of unconscious memory on my part. Also I most certainly did not know enough mathematics to know how to form an equation which would represent such a curve, or to know even of what type the equation must be. But I had come across such equations and drawn the curves represented by them; for instance, afterwards I found in my notebook the spiral $r\,\theta = \frac{1}{2}\,\pi\,a$, and the cardioid: $r = a\,(1 + \cos\theta)$.

"We had used no textbook, and in the full notes of the lectures I had attended, these were the two curves I found most similar to planchette's. If my brain produced the equation written by planchette, it must have been that I unconsciously formed an equation like some I had seen before, which by a curious coincidence chanced to represent a heart-shaped curve."

Commenting on this experiment, Sir Oliver Lodge writes:

"The equation which would naturally occur to anyone is the cardioid $r = a(1 + \cos\theta)$ and if this equation had been written by planchette there would have been nothing specially remarkable; for although not then in Mrs. Garrett Smith's mind, she had undoubtedly known it as a student.

"The equation written by planchette is not a familiar one and certainly would not be likely to occur to her, nor would it have occurred to me; but the sketch given does not profess to be an exact representation of the curve corresponding to the equation written by the planchette, but only represents her recollection of its general character.

"Mr. J. W. Sharpe, of Bournemouth, has been good enough to draw out an accurate graph of the curve and here is his drawing on a reduced scale:

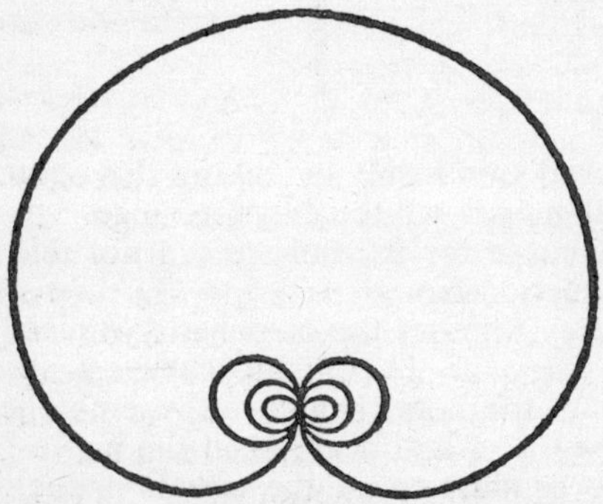

"With regard to his drawing Mr. Sharpe observes that the curve does not consist of two sets of spirals, as at first depicted roughly, but of two sets of loops, all passing through the cusp and touching one another there, and all contained within the outer heart-shaped boundary. The loops meet only at the cusp, and there is an infinite number of them.

"The equation very well represents the ordinary form of a planchette. But if it had accidentally been reversed into

$r = a\frac{\theta}{\sin\theta}$, the curve would have been entirely different

and entirely unlike any planchette outline.

"Mr. Sharpe thinks it very unlikely that either of the automatists had ever seen an accurate graph of the equation given in their writing. It is of course much more difficult to invent an equation to fit a given curve (which was the feat performed by the writing in this case) than, when the equation is given, to draw the curve represented by it."

Sometimes the unconscious intelligence, through bizarre figures and calculations, gives evidence of telepathy and premonition. This conclusion can be drawn from a series of surprising experiments (so surprising that they would appear absolutely incredible to anyone ignorant of the existence of paranormal phenomena) which were carried out in Brussels by a small psychic group led by M. Henri Poutet, a civil engineer and managing director of a large Belgian commercial firm, with the help of a medium who was not a professional but a distinguished member of the Brussels bar.

The dominant entity, which, apparently, communicated through the medium, called herself Stasia and claimed to be the "double" of a living person.[1]

Communication took place either by typtology (rapping), or by automatic writing. In this second case the medium writes the messages without his conscious will intervening, as if his hand were guided by a personality other than his own.

Here is the report of the group's first experiments.

The meeting was at 8 p.m. From 8 till 10 there was conversation about general matters. The meeting then adjourned to the room reserved for the experiments. Those present formed a "chain," that is they held one another by the hand, with the exception of M. Poutet who was taking notes. As is usual they were required to remain

[1] According to occulists, spiritualists and theosophists, living organisms, especially man, possess a sort of duplicate of themselves, called the "etheric double"; this is generally known by abbreviation as the "double." This subtle entity is supposed to be the seat of vital energy which in certain conditions, as during sleep, can escape provisionally from the material body. It is to be noted that this conception is very ancient: it is to be found in various forms in primitive animism.

absolutely passive, without knowing what sort of phenomena they were to be called upon to witness.

The entity Stasia gave her instructions, which were carried out by the members of the group.

Each in his turn, the experimenters drew eight cards at random from an ordinary pack. The first card turned up was the *ace of clubs* and the last the *nine of spades*.

Through the intermediary of the medium, M. Poutet then asked the entity Stasia:

"What are you going to do tonight?"

Stasia: "With the first card I am going to produce the last by a mathematical process."

M. Poutet: "How?"

Stasia: "Translate the first card into figures corresponding to the order of the letters of the alphabet."

This was done, with the following result:

A S T R E F L E (*as trèfle* = ace of clubs)

1 19 20 18 5 6 12 5

Stasia: "Divide this number by 125."

This was done, producing: 953,614,849.

Stasia: "Take out every other figure after the 9."

This was done, leaving 93,189.

Stasia: "Since with an ace I am going to make a 9, take out the 1 and replace it by a 9."

This was done, producing 93,989.

M. Poutet: "What must be done next?"

Stasia: "I will dictate to you by rappings the figures forming a certain number."

M. Poutet wrote down:

1,545,033,666,909,970,307

and asked Stasia: "What next?"

Stasia: "Multiply the number I have just dictated by the number obtained by the transformation of the ace of clubs; the nineteen figures to the left of the multiplication, which you will translate into letters corresponding to the alphabet, will give you the solution to the problem."

The following operation was then carried out:

```
        1545033666909970307
                      93989
    -----------------------
      1390530300218973 2763
```

```
        1545033666909970307
                      93989
   ------------------------
      13905303002189732763
     12360269335279762456
    13905303002189732763
   4635101000729910921
  13905303002189732763
  -------------------------
  145216169319201199184623
```

NEUF PICSTA SIA (*neuf pic Stasia* = nine of spades Stasia).

Describing and commenting on the experiment, M. Poutet writes: "The entity was no doubt proud of her prowess, since she signed the solution boldly. The participants, however, accustomed to experiments of this kind, could not help feeling a certain indefinable thrill, partaking of both satisfaction and uneasiness. An Inaudi would probably have been able to solve the multiplication rapidly, but what he would *not* have been able to do would have been, in the first instance, to induce the spectators to draw the two cards, ace of clubs and nine of spades, and in the second place to choose precisely those cards which alone permitted of the transformations necessary for Stasia to sign her work. Now in the group there was no conscious Inaudi; although the members were acquainted with mathematics, they were absolutely incapable of such a feat as this. And so the question arises: *unconscious mind* or *spirit?*—And if the hypothesis of the unconscious is admitted, does this unconscious intelligence possess the power to control the muscular system of the operators, thus abolishing their mastery of their own nervous systems and using them at will?

The following experiment, conducted according to the usual procedure, is at least as astonishing as that which has just been related.

M. Poutet drew a card at random and placed it, unseen by anybody, in the drawer of his desk. The participants, with the exception of M. Poutet, formed the chain.

M. Poutet said to Stasia:

"It is late. If you wish to operate by mathematics we would like you to act promptly."

Stasia: "Yes."

M. Poutet: "That suggests that you already know the card?"

Stasia: "Yes; write your Christian name and transform it into figures."

M. Poutet did this and obtained:

H E N R I
8 5 14 18 9

Stasia: "Multiply by a number composed of the figures which I will rap out and the fourteen last figures to the right will give the solution."

The following was written down:

11195215773166 2.

Stasia: "There is a mistake. 15 must be replaced by zero."

An argument arose and it was found in fact that Stasia was right: fifteen is not a *figure* but a *number*. But in that case why had fifteen been rapped out?

Stasia: "How can zero be rapped? With no sound? I have taken the letter O, the fifteenth."

The participants at last understood; they congratulated Stasia and declared themselves less intelligent than she.

Stasia: "Yes."

M. Poutet therefore replaced 15 by zero, did the multiplication and translated the answer into letters:

$$\begin{array}{r}
11195207731662 \\
8514189 \\
\hline
100756869584958 \\
89561661853296 \\
11195207731662 \\
44780830926648 \\
11195207731662 \\
55976038658310 \\
89561661853296 \\
\hline
95318114521631552118
\end{array}$$

NEUFCOEUR (*neuf coeur* = nine of hearts).

He then opened the drawer, took out the unknown card and turned it over. General astonishment: it was the *nine of hearts*.

It may be accepted in the light of these experiments, whether we consider those of the Société Universelle d'Etudes Psychiques, those of Oliver Lodge or those of the Belgian group (which last, let it be emphasized, are not unique of their kind), that lightning calculators can be otherwise inadequate mentally, sometimes even backward, and that the gift of calculation can make its appearance in childhood: a certain lack of balance, compared with the normal psychological state, between the subconscious faculties and the conscious faculties is even necessary to enable this power to express itself. Moreover as soon as the conscious faculties assert themselves too much the unconscious regions of the mind cease to manifest themselves; this can be observed among the few lightning calculators who have later become intellectuals.

At all events the problem of the lightning calculators undoubtedly shows that that which extends beyond the threshold and outside the limits of that part of our field of consciousness which is adapted to the needs of ordinary life, is at once more extensive and more complex than that which is included within those limits: this is the domain of the subconscious and psychic faculties of our being. At one end of this domain we find dreams, which are to some extent a normal product, of inferior quality and without great utility. At the other end we meet with special faculties and modes of cognizance, rarer, richer and more precious powers: telepathy, extrasensory perception and in general all the paranormal powers of man. Between these extremes are the intermediary virtues whose origin is the same but whose importance is very varied. These include the faculties of the calculators, painters, musicians and writers, which are associated more or less closely with the conscious faculties; they include also the extraordinary mnemonic powers we have found in Mlle. Osaka, and finally the faculties which express themselves in the minor medium-like manifestations of semi-intellectual character which constitute the usual stock-in-trade of spiritualist and psychic seances.

With Taine, therefore, we can "compare the mind of man to a theater of infinite depth, whose footlight rail is very narrow and whose stage widens out behind it. In front of this lit rail, there is hardly room for more than a single actor. Beyond, at the different depths of the stage, are other groups, less and less distinct as they are further from the footlights. Beyond these groups, in the wings and at the very back of the stage, are a multitude of dark shapes which a sudden call sometimes brings on to the stage or even at times right up to the footlights."

The footlights are the field of clear consciousness, where, as Taine observes, but one actor can perform. In contrast, on the different planes of the stage and in the penumbra of the backdrop and the wings, are gathered many actors whose faces and precise roles we do not know; they are the different forms of the unconscious.

CHAPTER II

Simulated Mental Calculation

THE MAJORITY of the arithmetical operations performed by lightning calculators can be simulated, that is replaced by others which resemble them only in appearance and are different in method.

The Magic Number

Among these exercises some deserve only brief mention for they would scarcely deceive even an untrained observer. These are the multiplications, divisions, additions, subtractions and various problems in which the operator directs the series of operations to be carried out or himself introduces one of the terms or factors. In these conditions one is tempted to suppose, as is indeed the case, that he knows the answers in advance, or, at least, that he obtains them very easily. This is precisely the case with the multiplication, by a multiplier between 2 and 200, of the "magic" number:

5 2 6 3 1 5 7 8 9 4 7 3 6 8 4 2 1 0.

At first sight this number appears like any other, but it has the curious property of giving products in which figures follow each other in a circular permutation, which enables the results of multiplications to be supplied instantaneously. You write it up carelessly on the blackboard, as if it were a number taken at random, turn your back to the board and ask spectators for a multiplier between 2 and 200. At once, reading from left to right, you give the answer.

Let us suppose that the multiplier in the first instance is less than or equal to 18. You make a cut in the multiplicand after the figure which is the same as the multiplier (or its 2nd figure if the multiplier has two figures) and next call out the figures which follow it, then the figures which precede the cut, and finally add a zero. Note however that except for the 9 and the 0 the figures repeat themselves twice. It is then necessary to take the following figure, choosing the lower when the multiplier lies between 2 and 9, or the higher figure when the multiplier lies between 11 and 18. Two examples will show the procedure at a glance.

Suppose that you have to multiply the number in question by 7.

The cut can be made between 7 and 8, or between 7 and 3. It is made this time between 7 and 3, because 7 lies between 2 and 9. You therefore write down:

3 6 8 4 2 1 0

then

5 2 6 3 1 5 7 8 9 4 7

to which zero is added, giving this final result:

3 6 8 4 2 1 0 5 2 6 3 1 5 7 8 9 4 7 0.

Now suppose that you have to multiply the magic number by 16. The cut can be made between 6 and 3 or between 6 and 8. This time it is made between 6 and 8, because the multiplier is between 11 and 18. You therefore write:

84210

and then:

5263157894736

which gives the product:

8421052631578947360.

Multiplication of the number by a multiplier lying between 19 and 200 is more complex, but also more surprising. Let us examine the different cases.

First of all for the multipliers 20, 30, 40, 50, 60, 70, 80, 90, it is sufficient to multiply the magic number by the figure of the tens, as would be done where the multiplier had only one figure, and add a second zero to the right of the answer. Obviously when multiplying by 100 two zeros are added at the end of the magic number, and this operation is therefore without interest.

Next, there are five multipliers which give special results. These are 19, 38, 57, 76 and 95.

The products are in fact composed only of 9 with three different figures: these are the first figure and the last two figures, the end figure being always a zero. Thus with 38 as the multiplier you get the figure 9 repeated 17 times, with 1 at the beginning, 8 as the figure of the tens and 0 for the figure of the units:

19999999999999999980.

In principle one should avoid using the special multipliers which reveal that the multiplicand is a number with peculiar characteristics. To this end one can have the two figures with which the multiplication is to be made drawn out of a hat. All the numbers from 2 to 200, except the numbers in question, will be written down on cards, and one of these selected at random.

For a multiplier lying between 21 and 29 inclusive, you first write down 1, then the magic number, as if it were multiplied by the 2nd figure of the multiplier increased by 1, making the cut at the lower succeeding figure, and reducing the last figure by one unit (which gives the 2nd figure of the multiplier); finally adding a zero.

Thus to multiply the number by 23, you proceed as follows. After having written 1, you write the result obtained by multiplying the magic number by $3 + 1 = 4$, the cut taking place before the lower following figure, in this case 2. You end with a 3 and a zero:

12105263157894736830.

If the multiplier is between 31 and 37 inclusive, the operation is identical with the preceding one, but the cut is made before the higher figure. Take for example 35 as the multiplier. You write 1 on the left, then the result obtained by multiplying the magic number by $5 + 1 = 6$, the cut taking place between 6 and 8. Finally you end with 5 and a zero, which gives:

18421052631578947350.

For a multiplier between 39 and 48 inclusive, you operate as with multipliers from 21 to 29, but you write 2 instead of 1 at the beginning of the answer, add 2 to the second figure of the multiplier and reduce the last figure but one by 2 units.[1]

Thus with 43 you obtain:

22631578947368421030.

If the multiplier lies between 49 and 56 inclusive the operation is still the same but the cut is made at the higher figure. If it is between 58 and 67 inclusive, the figure to write on the left is 3 and you write the magic number as if it were multiplied by the last figure of the multiplier increased by 3 units. The cut is made at the lower figure following and the last figure but one of the answer must be

[1] Note that you can obtain a number of two figures by adding 2 to the figure of the units of the multiplier. At this juncture only the last figure of the number obtained need be considered.

A similar observation applies to the cases which follow.

reduced by 3. For a multiplier between 68 and 75 inclusive the operation is the same as the preceding one but the cut is made at the higher figure. If the multiplier lies between 77 and 85 inclusive the cut is at the lower figure and if it is between 86 and 94, at the higher figure. In both cases the figure at the beginning is 4, as well as the additional figure. For a multiplier between 96 and 100, the cut is at the lower figure and the starting figure, like the additional figure, is 5. Finally, above 100 the operation is carried out in the same fashion: from 100 to 104 the procedure is identical with the preceding; the same for 105 to 113, but the cut is at the higher figure and so on. It should be noted however that the multipliers 114 and 152 give a series of 9. These are also numbers to avoid.

Instantaneous additions

The following numbers are written up on the blackboard:

1								
2	6							
3	9	15						
4	12	20	28					
5	15	25	35	45				
6	18	30	42	54	66			
7	21	35	49	63	77	91		
8	24	40	56	72	88	104	120	
9	27	45	63	81	99	117	135	153

By noting that each horizontal line forms an arithmetical progression whose ratio is twice the figure at the beginning, you can write these numbers very easily. Thus the numbers of the 6th line are obtained as follows:

Ratio of the progression:

$$6 \times 2 = 12.$$

Numbers of the progression:

$$\begin{aligned} &6 \\ 6 + 12 &= 18 \\ 18 + 12 &= 30 \\ 30 + 12 &= 42 \\ 42 + 12 &= 54 \\ 54 + 12 &= 66. \end{aligned}$$

It is the same for the other lines; thus for the last the ratio is $9 \times 2 = 18$ and you have successively:

$$9 + 18 = 27; 27 + 18 = 45; 45 + 18 = 63; \text{etc.}$$

When the numbers have been written down a spectator is asked to choose a horizontal line. Instantly you give the total of the numbers which compose it.

In reality no addition needs to be done, but you calculate the cube of the figure which starts the row. Thus the total of the 7th line is $7^3 = 7 \times 7 \times 7 = 343$; that of the 8th, $8^3 = 8 \times 8 \times 8 = 512$, etc.

In the mental calculation which follows, you can give the impression of doing additions and subtractions very rapidly and simultaneously.

You ask a spectator for any number, then for a number of two figures, another higher number and so on, the numbers being always higher and higher, all having two figures, or if need be, three figures.

These numbers are written by a spectator, one under another, without your seeing them, on a piece of paper or a blackboard.

At the same time he is asked to form a second column of numbers obtained by finding the difference between each number given and that which precedes it.

The calling out of the numbers is barely over when you supply the total of the two columns. The spectator can easily verify the accuracy of this extremely rapid mental calculation.

Example:

5	
17	12
23	6
34	11
55	21
69	14
85	16
98	13
386	93

The addition of the first column should be done mentally as the numbers are given. This is fairly easy, since, on the one hand they are numbers of two or three figures, and on the other hand the spectator who is occupied in writing up the numbers as well as their differences always leaves you the necessary time to carry out this small calculation.

To arrive at the total of the numbers in the second column is even simpler. It is only necessary to find the difference between the last number called and the first. In the example given you have:

$$98 - 5 = 93.$$

The following exercise is somewhat different.

A spectator is invited to write on the blackboard (without you seeing him) a series of consecutive numbers beginning with any number. When this has been done you ask how many numbers have been written down. If the answer is an uneven number, you ask for the number in the middle; if it is an even number for the two numbers in the middle or their sum. Immediately afterwards you give the sum of all the numbers written down.

All that is necessary is, in the first case, to multiply the middle number by the number of numbers which have been written, and in the second case to multiply the sum of the two middle numbers by half the number of numbers.

1st example.—*The spectator writes:*

13 14 16 17 18 19 20 21

making 9 numbers in all.

Nine being uneven, the middle number is asked for: 17.

The total of the numbers is therefore:

$$17 \times 9 = 153.$$

2nd example.—*The spectator writes:*

16 17 18 19 20 21 22 23

making 8 numbers in all.

Eight being even, you ask for the two middle numbers or their total: $19 + 20 = 39$.

The total of the numbers is therefore:

$$39 \times 4 = 156.$$

The total of the addition can also be given by asking only for the first and the last number of the series. This is how you proceed.

In the 1st example the spectator will reply 13 and 21. You add the total: $13 + 21 = 34$. The half of this supplies the number in the middle, 17, which you then multiply by the difference $21 - 13 = 8$, increased by 1. You are then led to the same result as before:

$$17 \times 9 = 153.$$

In the 2nd example the spectator supplies 16 and 23.

You find the total: $16 + 23 = 39$, then the difference: $23 - 16 = 7$, to which you add 1, which gives 8, which is then divided by 2. Thus you get:

$$39 \times 4 = 156.$$

Divining a number

The following arithmetical exercise, as well as others of the same kind, contains nothing to mystify anyone, and yet we have seen it carried out by an amateur "psychist" during a public performance of mental calculation. The calculator says to a spectator:

"Write down a number of three different figures; reverse it, that is write it backwards; then subtract the smallest number from the largest."

The spectator, secretly of course, writes for instance: 843 — 348 = 495. The calculator can then ask: "What is the first figure?", here 4, and immediately give the difference required: 495. He has only to subtract this first figure from 9 to have the last, 5; as for the middle figure, it is always 9. He can also ask nothing but continue as follows: "Reverse the difference obtained and add to it this new number, replacing the hundreds by zero, if the difference does not involve more than two figures." He at once gives the result: 1,089. In fact it is always this, whatever the initial number. Thanks to shrewd observation the illusionist Caroly has made the first part of this experiment much more mysterious and turned it into a real conjuring trick. The procedure consists in having the number written up on a blackboard or a piece of slate and listening to the noise made by the chalk, this being different for the ten figures of the numeration.

"The tracing of a straight line," writes Caroly, "produces a different sound from that for a curve. Moreover it will be noticed that there is a stop at each angle. Thus the figure 1 is formed of one straight line, which the ear soon learns to recognize. The 4 is made with three straight lines easy enough to perceive. The 2 makes a curved line, followed by a straight one. The 3 is formed of two loops, etc."

You therefore listen attentively to the sound produced by the tracing out of the first figure, then to that of the last figure when the spectator writes the number backwards. In this way the first impression can be checked. As we have seen, knowledge of this figure enables the required result to be determined.

Here is another exercise. The calculator says to a spectator:

"Think of a number; subtract 1, double the remainder; take away 1; add to what remains the number you first thought of; give the result."

Instantly the calculator gives the number first thought of! All he has to do is to add 3 to the result supplied by the spectator and then take the third of the sum.

Example: number thought of by the spectator, 6,457.

Operations carried out: 6,457 — 1 = 6,456; 6,456 +

6,456 = 12,912; 12,912 − 1 = 12,911; 12,911 + 6,457 = 19,368.

Operations carried out by the performer:

19,368 + 3 = 19,371; 19,371 ÷ 3 = 6,457.

The determination of a missing figure in the result of a subtraction or a multiplication is related to the previous exercises.

You have a number written down, then have another number formed with its significant figures arranged in any order. Then you have the smaller subtracted from the greater and ask for the difference, less one figure which must not be zero. At once you give the missing figure.

The principal of the experiment is that, in this particular case, the difference is always a multiple of 9. It is therefore sufficient to find the sum of the figures of the incomplete difference and to add to this a number which will make it a multiple of 9. This figure is the one which was not given:

$$\begin{array}{r} 85246 \\ 64258 \\ \hline 20988 \end{array}$$

If the spectator has removed the figure 8, he gives 2,098. Then you add: 2 + 0 + 9 + 8 = 19. The nearest multiple of 9 is 27. The missing figure is therefore 27 − 19 = 8.

A similar trick can be done with multiplication. You form a number by asking for a figure from each of six or seven spectators. A rapid calculation enables you to see if the number is a multiple of 9. If it is not you surreptitiously add the necessary figure. As the number of figures asked for is fairly large the subterfuge passes unnoticed. A spectator is asked to multiply the number formed by a number he chooses, then to give the result with one figure omitted, a condition being that this should not be a zero. The figures of the difference, less one figure, can be furnished in any order. In this case, as in the other, you immediately give the missing number.

Here again it is sufficient to add up the significant figures of the incomplete difference and add the figure needed to make it a multiple of 9. This is the figure required.

The exercise which we are now going to describe is a little more subtle than the previous ones. You ask a spectator to take 9 coins, or failing this 9 matches, and to put 3 of them in one hand and 6 in the other. You then make him do the following mental calculations:

1.—Double the number of coins which are in your right hand; 2.—treble the number in your left hand; 3.—add the two answers obtained; 4.—divide this sum into two equal parts; 5.—take 11 away from one of these two halves; 6.—double the remainder and add 1.

At this moment, without the spectator having supplied the result of the operations, you state the number of coins held in his right hand.

In fact the spectator unknowingly gives the information necessary for the success of the experiment. The 4th and 5th operations are indeed only possible if he has put 3 coins in the right hand and 6 in the left. In this case he carries out the calculations without making any remark and you announce that he has 3 coins in the right hand, which is correct. If he has put the 3 coins in his left hand he will generally say when he comes to the fourth operation that the division is impossible. At this moment the contents of his hands will be known. If he makes no remark and carries out the division, which sometimes happens, he will obtain 10.5. But at the fifth operation he will point out that this cannot be done because it would be necessary to take away 11 from 10.5. You then say approximately this:

"That does not matter. Take away 9 and multiply the result by 2. You will then have the number of coins which you are holding in your left hand."

And in fact the operation gives: $1.5 \times 2 = 3$.

The number of Destiny

Finally, in this series of calculations, comes an exercise of a somewhat different kind which is also rather original.

The performer turns to a spectator and says:

"Without knowing anything about you I am going to give you the number of your destiny, that is, the number which is dominant in your life. It is 3,918. When were you born?" The spectator may say, "1904." "And your wife?" "1914." "Now, if you write down those two dates and your respective ages at your last birthdays—that is, 55 and 45—and add them all together, you will find 3,918. This number is that which dominates your whole existence since it is formed with dates which are important to you."

In reality this number would have been the same for any other person. It is the double of the current year, here

1959. If a third person also asks for his destiny number he can be given the triple of the current year, in which case it will be necessary to add the date of birth and the age of one of his children.

Rapid multiplication

Having dealt with these few exercises, we come now to arithmetical operations which by the rapidity of their execution and sometimes by their complexity, recall those carried out by the lightning calculators.

The most interesting are concerned with multiplications and the extraction of roots, tricks with the calendar and memory for numbers.

Where the multiplications are concerned, let us examine certain procedures enabling some simple products to be found rapidly.

1. *Multiplication of two numbers between 10 and 20.*

To one of these numbers you add the units of the other, then to this sum, considered as tens, you add the product of the units.

1st example.—*To obtain the product of:*

$$12 \times 18$$

You calculate mentally:

$$12 + 8 = 20 \text{ tens}$$
$$2 \times 8 = 16$$

Product required:

$$20 \text{ tens} + 16 = 216.$$

2nd example.—*To obtain the product of:*

$$15 \times 19$$

You calculate mentally:

$$15 + 9 = 24 \text{ tens}$$
$$5 \times 9 = 45$$

Product required:

$$24 \text{ tens} + 45 = 285.$$

Note.—In these calculations, as in those which follow, the words tens and hundreds are ignored.

2. *Multiplication of two numbers of two figures of which the tens are the same.*

You add to one number the units of the other and multiply this sum by the figure of the tens. To the product, considered as tens, you add the product of the units.

1st example.—*To obtain the product of:*

$$46 \times 49$$

You calculate mentally:

$$46 + 9 = 55$$
$$55 \times 4 = 220 \text{ tens}$$
$$6 \times 9 = 54$$

Product required:

$$220 \text{ tens} + 54 = 2{,}254.$$

2nd example.—*To obtain the product of:*

$$85 \times 87$$

You calculate mentally:

$$85 + 7 = 92$$
$$92 \times 8 = 736 \text{ tens}$$
$$5 \times 7 = 35$$

Product required:

$$736 \text{ tens} + 35 = 7{,}395.$$

Note.—If the figure of the tens is 9, the procedure is the following. You find the figure which must be added to each of the factors to make up 100 and you multiply together the two figures so obtained; the result represents the two last figures of the final product. The two first figures of the latter are obtained by taking away from the multiplicand the figure which was lacking to the multiplier to make 100, or vice versa.

Example.—*To obtain the product of:*

$$95 \times 98$$

You calculate mentally:

$$95 + 5 = 100$$
$$98 + 2 = 100$$
$$5 \times 2 = 10$$
$$95 - 2 = 93 \text{ or } 98 - 5 = 93$$

Product required:

$$9{,}310.$$

3.—*Multiplication of two numbers of two figures ending with the same figure.*

To the product of the tens, considered as hundreds, you add the product, considered as tens, of the sum of the tens and the figure of the units, then adding the square of the units.

1st example.—*To obtain the product of:*

$$94 \times 54$$

You calculate mentally:

$$9 \times 5 = 45 \text{ hundreds}$$
$$9 + 5 = 14$$
$$14 \times 4 = 56 \text{ tens}$$
$$45 \text{ hundreds} + 56 \text{ tens} = 506 \text{ tens}$$
$$4 \times 4 = 16$$

Product required:

$$506 \text{ tens} + 16 = 5{,}076.$$

2nd example.—*To obtain the product of:*

$$72 \times 32$$

You calculate mentally:

$$7 \times 3 = 21 \text{ hundreds}$$
$$7 + 3 = 10$$
$$10 \times 2 = 20 \text{ tens}$$
$$21 \text{ hundreds} \times 20 \text{ tens} = 230 \text{ tens}$$
$$3 \times 3 = 9$$

Product required:

$$230 \text{ tens} + 9 = 2{,}309.$$

4. *Multiplication of two numbers of two figures ending with 5.*

To the product of the tens, considered as hundreds, you add half the sum of the tens, then add 25 to the result.

Example.—*To obtain the product of:*

$$35 \times 75$$

You calculate mentally:

$$3 \times 7 = 21 \text{ hundreds}$$
$$\frac{3 + 7}{2} = 5$$
$$21 + 5 = 26 \text{ hundreds}$$

Whence the product required:

2,625.

When the sum of the tens is an odd number, you take the half of it to the nearest whole number below and add 75 instead of 25.

Example.—*To obtain the product of:*

$$35 \times 65$$

You calculate mentally:

$$3 \times 6 = 18 \text{ hundreds}$$
$$\frac{3 + 6}{2} = 4.5$$

You retain only 4.

$$18 + 4 = 22 \text{ hundreds}$$

Whence the product required:

2,275.

5. *Square of any number ending in 5*

The 5 is disregarded and the number remaining is multiplied by that which immediately follows it numerically; then 25 is added to the product.

1st example.—*To find the square of:*

85

You calculate mentally:

$8 \times 9 = 72$

Whence the square of 85:

7,225.

2nd example.—*To find the square of:*

345

You calculate mentally 34 × 35 by the procedure given in (2). You obtain 1,190. The square of 345 is therefore:

119,025

6. *Multiplication performed without writing down the part products.*

The procedure is not strictly speaking one of purely mental calculation, but it enables the product of two numbers to be found very quickly.

(A) *Where the multiplicand and the multiplier have two significant figures.*

Suppose 78 is to be multiplied by 83. The operation is set out in the usual fashion, thus:

$$\begin{array}{r} 78 \\ 83 \\ \hline 6{,}474 \end{array}$$

And the following mental calculations are carried out:

(a) 8 × 3 = 24; you put 4 under the units and carry 2.

(b) You find the sum of the products of the four unit figures diagonally, thus 7 × 3 = 21; 8 × 8 = 64; total 85; you add the figure carried 2, which gives 87; you put 7 in the place of the tens and carry 8.

(c) Finally you find the product of the two remaining figures, 7 × 8 = 56, to which you add the 8 you have carried, yielding 64; you put 64 to the left of the figure of the tens and obtain the final product: 6,474.

Note.—It is best to add each of the carried figures immediately. Thus you add 2 to 21, which gives 23 and you calculate: 23 + 64 = 87.

(B) *Where the multiplicand and the multiplier have 3 significant figures.*

Suppose 695 to be multiplied by 327. The problem is set out:

695
327

227,265

And, as before, the following calculations are made:

(a) 5 × 7 = 35; you put 5 under the units and carry 3.

(b) 9 × 7 = 63 + the figure carried, 3, = 66; 5 × 2 = 10; total 76; put down 6 in front of 5 and carry 7.

(c) 6 × 7 = 42 + the figure carried, 7, = 49; 5 × 3 = 15; total 64; 9 × 2 = 18; 64 + 18 = 82; put down 2 and carry 8.

(d) 6 × 2 = 12 + the figure carried, 8, = 20; 9 × 3 = 27; 20 + 27 = 47; put down 7 and carry 4.

(e) 6 × 3 = 18 + the figure carried, 4, = 22; put down 22.

The answer required is 227,265.

Note.—If the multiplier has only two figures, you can replace the figure of the hundreds by a zero. A similar remark can of course be made in connection with any multiplication.

Suppse 735 is to be multiplied by 28. The problem is set out:

735
028

20,580

And you carry out the following mental calculations:

(a) 5 × 8 = 40; you put 0 under the units and carry 4.

(b) 3 × 8 = 24 + the figure carried 4, = 28; 5 × 2 = 10; total 38; put down 8 and carry 3.

(c) 7 × 8 = 56 + the figure carried, 3, = 59; 5 × 0 = 0; 3 × 2 = 6; total 65; put down 5 and carry 6.

(d) 7 × 2 = 14 + the figure carried, 6, = 20; 3 × 0 = 0; total 20; put down 0 and carry 2.

(e) 7 × 0 = 0 + the figure carried, 2, = 2; put down 2 and obtain the final product: 20,580.

(C) *Where the multiplicand and the multiplier have 4 significant figures.*

Suppose 8,673 is to be multiplied by 4,589. The problem is set out:

8,673
4,589

39,800,397

And you carry out the following mental calculations:

(a) 3 × 9 = 27; you put 7 under the units and carry 2.
(b) 7 × 9 = 63 + the figure carried, 2, =65; 3 × 8 = 24; total 89; put down 9 and carry 8.
(c) 6 × 9 = 54 + the figure carried, 8, = 62; 3 × 5 = 15; 7 × 8 = 56; total 133; put down 3 and carry 13.
(d) 8 × 9 = 72 + the figure carried, 13, = 85; 3 × 4 = 12; 6 × 8 = 48; 7 × 5 = 35; total 180; put down 0 and carry 18.
(e) 8 × 8 = 64 + the figure carried, 18, = 82; 7 × 4 = 28; 6 × 5 = 30; total 140; put down 0 and carry 14.
(f) 8 × 5 = 40 + the figure carried, 14, = 54; 6 × 4 = 24; total 78; put down 8 and carry 7.
(g) 8 × 4 = 32 + the figure carried, 7, = 39; put down 39.

The required product is 39,800,397.

(D) *Where the multiplicand and the multiplier each have any number of significant figures.*

When the multiplicand and the multiplier each have more than 4 figures, the mechanism of the operation does not change, but becomes rather complicated. Thus, to multiply 238,756 by 12,321, the following calculations have to be made:

(a)	6×1 =	6
(b)	(5×1)+(6×2) =	17
(c)	(7×1)+(6×3)+(5×2) =	35
(d)	(8×1)+(6×2)+(7×2)+(5×3) =	49
(e)	(3×1)+(6×1)+(8×2)+(5×2)+(7×3) =	56
(f)	(2×1)+(3×2)+(5×1)+(8×3)+(7×2) =	51
(g)	(2×2)+(3×3)+(7×1)+(8×2) =	36
(h)	(2×3)+(3×2)+(8×1) =	20
(i)	(2×2)+(3×1) =	7
(j)	2×1 =	2
		2941712676

Extraction of square roots

For the extraction of a square root the mental calculation broadly follows the normal method: you divide the number into groups of two figures, you look for the largest square contained in the first group, you carry the root as the first result and subtract this square from the first group.

Next, you mentally add on to the remainder the first figure of the second group and divide the number thus obtained by twice the first figure carried; you retain this quotient as the second figure of the root required. As can be seen, the process is rather complicated and lengthy. However you can simplify it considerably, especially when it is applied to the roots of perfect squares having only a small number of significant figures.

1. *The number given, a perfect square, has 4 figures and the root has 2 figures.*

You learn by heart the squares of the 9 figures:

Numbers	*Squares*
1	1
2	4
3	9
4	16
5	25
6	36
7	49
8	64
9	81

Except for 5, two possible values for the root correspond to each final figure of the squares. Thus, to the final figure 4, 2 or 8 correspond; to 9, 3 or 7 correspond, etc.

1st example.—*To calculate the square root of 3,136:*

The number is split into groups of two figures:

31 36.

The terminal figure of the right-hand section indicates, according to the table, that the probable figure of the units of the root is 4 or 6.

You then look for the perfect square immediately inferior to the group on the left, which supplies the figures of the tens. This perfect square is 25, of which the root is 5. The root required is therefore 54 or 56.

Its actual value will be fixed by the value of the product:

$$5(5+1) = 5 \times 6 = 30.$$

This product being less than 31, the root wanted is the greater, that is to say 56.

2nd example.—*To calculate the square root of 5,329.*

You separate the number into groups of two figures:

53 29.

Figure of the units: 3 or 7.

Square immediately inferior to 53: 49.

Root of 49: 7.

Root required: 73 or 77.

The product: $7(7 + 1) = 7 \times 8 = 56$ being *greater* than 53, it is the smaller root which must be retained, i.e., 73.

2. *The number given, a perfect square, has 5 or 6 figures and the root has 3 figures.*

Two cases must be considered:

(A) *The number given is less than 40,000.*

Besides the previous table, that of the perfect squares from 10 to 20 also needs to be known:

A	*B*
10	100
11	121
12	144
13	169
14	196
15	225
16	256
17	289
18	324
19	361
20	400

1st example.—*To calculate the square root of 24,649:*

First take the group of two figures to the right of the number, i.e., 49.

The terminal figure of this number, according to the first table, indicates that the figure of the units of the root is 3 or 7. It will be seen in column *B* of the second table that 246 comes immediately after 225 and before 256. The number of the tens of the root is therefore 15. And the root will be 153 or 157.

To fix the figure of the units you find the product:

$$15\ (15 + 1) = 15 \times 16 = 240.$$

240 being *less* than 246, it is the *larger root* which fits, that is 157.

2nd example.—*Root of 26,569?*

Figure of the units: 3 or 7.

Square immediately inferior to 265: 256.

Root of 256: 16.

Root required: 163 or 167.

The product $16\ (16 + 1) = 16 \times 17 = 272$ being greater than 265, it is the *smaller root* which fits, that is 163.

(B) *The number given is greater than 40,000.*

You split the number into groups of 2 figures starting from the right; you determine the figure of the hundreds of the root and proceed thereafter as in the written operation.

1st example.—*To extract the square root of 191,844.*

Divide the number into groups of two figures starting from the right:

19 18 44.

You consider the third section from the right, that is 19. The square which is immediately inferior is 16, of which the root is 4. This is the figure of the hundreds of the root required.

You find the difference:

$$19 - 16 = 3$$

You put this 3 before the left-hand figure of the second group, which gives 31; this is the dividend.

You double the 4 and add 1, which gives 9; this is the divisor.

You look for the nearest quotient, above or below, which will yield the *minimum remainder*. In this example you obtain 3, which is the quotient nearest below. On consulting the first table it is noticed that, according to the first group, 44, the figure of the units is 2 or 8. The quotient 3 being the nearest below, you choose the *larger root,* that is 438.

2nd example.—*To extract the square root of 58,564.*

Separate this number into groups of two figures starting from the right:

5 85 64.

The square immediately inferior to 5 is 4.

Its root is 2.

Difference: $5 - 4 = 1$.

This 1 is placed before the 8, which gives 18 and this number is then divided by $2 \times 2 + 1 = 5$, which supplies 4 as the nearest above. The figure of the hundreds is 4.

According to the first table, 2 or 8 corresponds to 64. As 4 has been obtained as nearest above, the *smaller root,* i.e., 242, is chosen.

Notes.

(1) If the division is made without remainder, the smaller root is also chosen.

(2) When the square ends in a 6, the significant figure of the units of the root is 4 or 6, and, in this case, the rule of nearest above or below does not in general enable a choice to be made between the small and the large root.

Before concluding a check must be made by 9 or 11, a circumstance which obviously complicates the resolution of the problem.

Extraction of perfect cube roots and higher roots

Let us look now at some methods for the rapid extraction of cube roots and higher roots.

In respect of cube roots we shall first examine a procedure limited to the extraction of roots of two or three figures—corresponding already to cubes of "impressive" size, since they can be formed from 8 or 9 significant figures.

1.—*Roots of two figures.*

To find these instantaneously the table of cubes for the first 9 numbers must be known, i.e.:

Numbers	*Cubes*
1	1
2	8
3	27
4	64
5	125
6	216
7	343
8	512
9	729

It will be seen at once that the figure of the units of the cubes is the same as the root for 1, 4, 5, 6 and 9, and that for the other figures there exists a criss-cross relationship by which the end figure can be easily remembered.

The cube of 2 is 8.

The cube of 8 ends in 2.

The cube of 3 ends in 7.

The cube of 7 ends in 3.

The figure of the *units* of the cube root is thus given immediately by the end figure of the cube concerned. It is in fact the same figure if the cube ends in 1, 4, 5, 6 and 9; it is equal to 2 if the figure of the units of the cube is 8, equal to 3 if the end figure of the cube is 7, and inversely.

To find the second figure of the root, i.e., that of the *tens*, the same table is used, but in a different way.

The number to be dealt with is split into groups of 3 figures, starting from the right, and only that of the thousands is retained. You refer mentally to the table, where you find this number sandwiched between two consecutive cubes. The smaller of these two is chosen and its root is the figure of the tens of the root required.

An example will show how this works.

Suppose that the cube root of 658,503 is to be determined.

As the number ends in 3, the figure of the units of the root is 7, since the cube of 7 ends in 3.

The group of the thousands: 658, comes between 512 and 729 on the table. The figure of the tens of the root corresponds to 512. It is therefore 8.

Therefore the cube root required is 87.

2.—*Roots of three figures.*

To extract roots of 3 figures, whose perfect cubes can in consequence have 7, 8 or 9 significant figures, you must know the following table in addition to the first table:

A	*B*
1	1
2	7
3	9
4	5
5	3
6	8
7	6
8	2
9	4
10	10

This table being known, let us suppose that the cube root of a perfect cube is to be extracted. The number is split into groups of 3 figures starting from the right, but the group at the extreme left may have only one or two figures. The figure of the *units* (u) and that of the *hundreds* (h) are determined as if for a root of two figures, and consequently the first table is used.

It remains to find the figure of the *tens*. For this you find the sum of all the "odd" (first, third, fifth and seventh) figures of the cube, working from right to left; next you add up "even" (second, fourth, sixth and eighth) figures in the same way; finally you subtract the sum of the "evens" from the sum of the "odds" (if the "evens" total is greater than the "odds," add 11 to the "odds"). The result of the operation is a figure (*a*). You now refer to the second table. In column *A* will be found the figure you have just obtained. You note the corresponding figure in column *B*. This is the figure which will enable you to find rapidly the figure of the tens required. It is sufficient to add *h* and *u* and subtract from the total the figure given in *B*. The result is the figure of the *tens*. Here again if *B* is greater than $h + u$, you add 11 to this total and then do the subtraction.

Two examples will show the sequence of calculations to be done.

Suppose that the cube root of 14,886,936 is to be found. The figure of the *units* and that of the *hundreds* are obtained by means of the first table. The figure of the units is 6, since the number given ends in 6. That of the hundreds is 2, since 14 is sandwiched between 27 and 8. It remains to determine the figure of the *tens*. You add up the first series of "odd" figures, counted from the right:

$$6 + 9 + 8 + 4 = 27.$$

Next you find the sum of the series of "even" figures:

$$3 + 6 + 8 + 1 = 18.$$

You subtract the second total from the first:

$$27 - 18 = 9.$$

You now refer to the second table. Corresponding with 9 in column *A* you find 4 in column *B*. You find the sum of the figure of the units and of that of the hundreds already obtained:

$$6 + 2 = 8.$$

From this total you take away 4, which gives 4 for the figure of the *tens*.

The cube root required is therefore 246.

Another example: You are given the number 48,627,125 of which the cube root is to be determined. The figures of the *units* and of the *hundreds* are obtained immediately. They are respectively 5 and 3. You find the sum of the "odd" figures counting from the right:

$$5 + 1 + 2 + 8 = 16.$$

Then you find the total of the "even" figures:

$$2 + 7 + 6 + 4 = 19.$$

As 16 is inferior to 19, you add 11 to it, which gives 27. You subtract 19 from 27, which gives 8. You refer back to the first table. Corresponding to 8 in column *A* you have 2 in column *B*. You make the total of the figures of the units and the tens previously obtained, that is $5 + 3 = 8$ and subtract 2; this leaves 6. This is the figure of the *tens*. The cube root required is therefore 365.

All these operations which at first sight appear rather complicated are in reality very easy to do with a little practice. The sole difficulty consists in learning the two lists of figures by heart, but if you do not wish to give yourself this trouble, and are working on a blackboard, you can perhaps write the lists beforehand on your cuff. A glance is enough to give you the indication needed at an opportune moment.

For the second method for the extraction of cube roots which we now give, and which moreover presents numerous points of similarity with the previous method, we are indebted to René Quinton. Here it is as set out by the eminent biologist:

"First of all you learn the following series of numbers: 1, 8, 27, 64, 125, 216, 343, 512, 729, 1 million. The nine first figures in fact stand for thousands: 1,000, 8,000, 27,000, 64,000, etc. Count on your fingers while you repeat 1, 8, 27, etc., to yourself. In this way you will obtain the figure of the tens of the cube roots. Thus, suppose that I have to find the cube root of 13,384. On hearing 13,000 I count on my fingers 1, 8, 27. The number 13 (of 13,000) lies between 8 and 27, that is between the second and third fingers. The figure of the tens of my root will then be 2. I listen to the end of the reading of the number 13,384. I remember the last figure, 4, and I say 24; 24 is the root asked for. The method will be the same for all numbers I am given which end in the figures 0, 1, 4, 5, 6, 9.

"Examples:

Perfect cube	*Cube root*
85,184	44
287,496	66
753,571	91
166,375	55

"For numbers ending with the figures 2, 3, 7 or 8, the figure of the units of the cube root is no longer the terminal figure, but it is still easily obtained. We subtract our terminal figure from 10. A final figure 2 thus gives us 8; a final figure 3 gives us 7; 7 gives us 3; 8 gives us 2. These figures 8, 7, 3, 2 (all complementary of 10) are the figures of the units of our roots.

"Examples:

Perfect cubes	*Cube roots*
10,648	22
54,872	38
148,877	53

"Below 1,000 the given number will have a cube root of one figure which can be found instantaneously by the same method.

"Examples:

Perfect cubes	*Cube roots*
64	4
125	5
216	6
8	2 (that is $10 - 8 = 2$)
27	3 (that is $10 - 7 = 3$)
343	7 (that is $10 - 3 = 7$)

"Above 1 million it is sufficient to establish points of reference for the cubes of 110, 120, 130, 140, etc., by simplifying these.

"It should be added that this method for the extraction of cube roots can be employed without much modification for finding seventh, eleventh, fifteenth, nineteenth roots, etc."

Instantaneous extraction of a fifth root

The method has also been devised by René Quinton.

"Memorize the following numbers," he writes, "1, 3, 24, 100, 300. Bear in mind that 1 stands for 100,000 and that the four numbers after that stand for millions, 3 millions, 24 millions, 100 millions, 300 millions; count on your fingers while you are repeating 1, 3, 24, 100, 300, and you will know already how to extract the fifth root of numbers having roots between 1 and 50.

"Example: I ask you to extract the fifth root of the number 4,084,101. I read this number to you slowly. As soon as I have said '4 million' you count on your fingers repeating one after another the numbers 1, 3, 24; you note that '4 million' lies between 3 and 24, that is between the second and the third fingers. Retain the figure 2 of this second finger; this is the figure of the tens of the root you are looking for. It remains only for you to hear me read out to the end the number 4,084,101. Pay no attention to the figures you hear except the last, that is the figure 1. As soon as you have heard it, add it to the right of the figure 2 of the tens, and say aloud: 21; this is the fifth root asked for.

"Let us go further. Now that you know the series 1, 2, 3, 24, 100, 300, learn the following series, still counting on your fingers: 777 millions, 1,500 millions, 3,000 millions, 6,000 millions, 10,000 millions. You will now know how to extract the fifth root of numbers whose root is between 1 and 100.

"Example: Extract the fifth root of 8,587,340,257. As soon as I have said '8,000 millions' you count on your fingers, saying over to yourself 777 millions, 1,500 millions, 3,000 millions, 6,000 millions, 10,000 millions. You perceive that 8,000 millions lies between 6 and 10,000 millions, or between the ninth and tenth fingers. Retain the figure 9 of this ninth finger. Listen to me reading out to the end the number 8,587,340,257. Pay no attention until the last figure 7. Add 7 to the right of the 9, this gives you 97. Say aloud 97; again, this is the fifth root asked for.

"We can go further still. Count on your fingers: 16, 24, 37, 53, 75. Tell yourself that these numbers stand for thousands of millions. You know how to extract the fifth root of numbers whose root falls between 1 and 150.

"Example: I ask you to extract the fifth root of 64,097,-340,625. As soon as I have said '64 thousand millions,' count on your fingers: 16, 24, 37, 53, 75. You see that 64 is situated between 53 and 75, that is between the fourth and fifth fingers. Retain the figure 4 of this fourth finger, and add 10 to it, which gives 14. Listen to the end of the reading of the number 64,097,340,625. As soon as I have uttered the last figure of this number, that is the figure 5, add it to the right of 14, this gives you 145, which is the fifth root asked for.

"In short, a fifth root can be extracted instantaneously. Very simple reference points (1, 3, 24, 100, 300, 777, etc.) supply the figures of the tens; the last figure of the number whose root is sought being that of the units.

"Here are some examples:

Perfect power	*Fifth root*
537,824	14
39,135,393	33
345,025,251	51
229,345,007	47
8,587,340,257	97

"When the number given is less than 100,000," notes M.

Quinton, "the figure of the tens is nil; the root is represented by the last figure of the number."

Examples: 3,125; 59,049; 32; corresponding fifth roots: 5; 9; 2."

Extraction of ninth, thirteenth, seventeenth and twenty-first roots

The extraction of these roots is carried out as for the fifth root, except that the reference points for the tens are altered. The figure of the units is always the figure of the original number given.

While there is no doubt that these methods enable apparently very difficult operations to be solved almost instantaneously, it should be borne in mind that they apply in general only to perfect powers and that they are useless for square roots of more than two figures, or for sixth roots, etc. When it has been once again emphasized that there are no simple methods of carrying out rapid subtractions, additions, multiplications and divisions which do not have a special character or in which the calculator has not himself introduced one of the factors, it will be acknowledged that the lightning calculator retains a clear superiority over these calculation devices.

He can however be equalled or even surpassed by the illusionist operating on a stage and having at his disposal a portable calculating machine, of small compass yet capable of carrying out complex operations in a few seconds: multiplications with multiplicand and multiplier of 8 or 10 figures or more, divisions of the same kind, extraction of square and cube roots of 4, 5, 6 figures, etc. All that is then necessary is that an assistant stationed in the wings should note the numbers given by the spectators, work the machine and transmit the results to the performer through a speaking tube, a telephone or a miniature short wave radio set. The use of a bandage, intended to conceal earphones, can be explained away by the necessity for "psychical isolation," or to show that there is no deception! It is easy to imagine still further means of transmission. Thus the assistant hidden in the wings can simply write the results on a blackboard, and show them to the operator without the knowledge of the audience.

The calendar demonstration

The calendar demonstration, which usually forms part of a stage mental calculation performance and which consists of finding the day of the week corresponding to a given date, can be carried out as easily as the previous calculations, by means of several methods (Albertini, Dhotel, Monthenolle, etc.) all based in general upon the same principle. Let us give the simplest here and apply it first of all to the period extending from 1860 to the present day, for this is generally the only one likely to interest an audience.

Three tables have to be memorized.

The first involves the figure 7 and its first three multiples: 7, 14, 21, 28.

The second table is this:

January	corresponds	to	0
February	”	”	3
March	”	”	3
April	”	”	6
May	”	”	1
June	corresponds	to	4
July	”	”	6
August	”	”	2
September	”	”	5
October	”	”	0
November	”	”	3
December	”	”	5.

The third table is a little more complicated:

1860 and 1900	correspond	to	0
1864 and 1904	”	”	5
1868 and 1908	”	”	3
1872 and 1912	”	”	1
1876 and 1916	”	”	6
1880 and 1920	”	”	4
1884 and 1924	”	”	2
1888 and 1928	”	”	0
1892 and 1932	”	”	5
1896 and 1936	”	”	3
1940	”	”	1

In the calculations which follow, it is further necessary to bring the numbers down to a figure lower than 7, starting from the principle that 7 and its multiples must first be deducted. Thus, 8 is brought down to 1, since $8 - 7 = 1$;

16 is brought down to 2, for 16 — 14 = 2; 27 is brought down to 6, since 27 — 21 = 6.

This principle being borne in mind and the tables learnt, three operations are necessary to obtain the desired result.

The first operation consists of transforming the day of the month into a figure inferior to 7, according to the principle we have just set out.

The second operation consists of adding to the figure obtained a figure corresponding to the month and supplied by the second table. The total will be brought down to a figure below 7 according to the method already employed.

Finally, in the third operation, this figure will be added to that given in the third table.

Two cases now have to be considered.

First case.—If the year chosen is a leap year the figure is furnished by the table. However, if the month asked for is January or February, it is necessary to reduce this figure by one unit. When the addition has been done, the total will be brought down below 7 in conformity with the general rule. The figure so obtained will supply the day of the week, it being accepted that 1 corresponds to Monday, 2 to Tuesday, 3 to Wednesday, etc., and 0 to Sunday.

It should be noted that 1900, which is divisible by 4 and which ought to be a leap year, is not, because it is a centennial and centennials are leap years only every four hundred years. There will be no need, therefore, to reduce by one unit for January and February, 1900.

Second case.—If the year chosen is not a leap year, the nearest year of earlier date will be considered, the figure which corresponds to it will be retained and the difference between the two years added to this figure. If the total is higher than 7, this figure will be subtracted. The answer will be added, as before, to that furnished by the two first operations, which will enable the required day of the week to be determined.

A few examples will show the simplicity of the method.

1st example.—*What day was June 9, 1884?*

First operation.—9 must be brought down to a figure lower than 7. We subtract 7, which leaves 2.

Second operation.—The figure corresponding to June is 4. We count 2 + 4 = 6. This total being lower than 7, we retain it.

Third operation.—1884 appears in the table and the figure which corresponds to it is 2. We therefore have: 6 + 2 = 8. This figure being higher than 7, we subtract 7, which leaves 1. Therefore the day of June 9, 1884 was a Monday.

2nd example.—*What day was January 4, 1884?*

First operation.—4 being lower than 7 there is no need to transform it.

Second operation.—The figure corresponding to January is 0. We therefore have: $4 + 0 = 4$.

Third operation.—The figure which corresponds to 1884 is 2. But as the month asked for is January, it is necessary to reduce this figure by one unit. We have thus: $4 + 1 = 5$. The day of January 4, 1884 was consequently a Friday.

3rd example.—*What day was June 5, 1898?*

First operation.—5 being lower than 7, there is no need to transform it.

Second operation.—The figure corresponding to June is 4. We count $5 + 4 = 9$. This total being higher than 7, we take this figure from it, which gives 2.

Third operation.—1898 does not appear in our third table. The nearest year of earlier date is 1896. The figure which corresponds is 3, to which we add the difference $1898 - 1896 = 2$. We obtain 5 and we add the 2 calculated in the second operation, which gives us 7. From this figure we subtract 7 and have 0. The day of June 5, 1898 was therefore a Sunday.

4th example.—*What day was March 27, 1900?*

First operation.—27 must be brought down to a figure below 7. We therefore take away 21, which leaves 6.

Second operation.—March corresponds to 3. We count: $6 + 3 = 9$, from which we subtract 7, leaving us with 2.

Third operation.—As 1900 corresponds to 0, the day of March 27, 1900 was a Tuesday.

5th example.—*What day was November 11, 1918?*

First operation.—Bring 11 down to a figure lower than 7: $11 - 7 = 4$.

Second operation.—November corresponds to 3. We therefore have: $4 + 3 = 7$. We bring this total down to a figure lower than 7: $7 - 7 = 0$.

Third operation.—1918 does not appear in our third table. The nearest year of earlier date is 1916, for which the figure is 6. Hence: $0 + 6 = 6$. To this result we add the difference $1918 - 1916 = 2$, which gives us 8, from which we subtract 7, which gives us 1. The day of November 11, 1918 was therefore a Monday.

Let us note that 1860, 1888 and 1928, all leap years, have a value of 0. If asked for a date in January or February of these years we replace 0 by 7, since these two figures are equivalent, and calculate: 7 — 1 = 6.

Moreover, because of the special value of 1940, which is 1 instead of 0, one needs to note that the figures which correspond to the years coming after that date are: 2 for 1941, 3 for 1942, 4 for 1943, 6 for 1944, 4 for 1948, 2 for 1952, and, according to the general rule, 7 or 0 for 1945, 8 or 8 — 7 = 1 for 1946, 9 or 9 — 2 = 7 for 1947, 5 for 1949, 6 for 1950, 7 or 0 for 1951, 3 for 1953, 4 for 1954, 5 for 1955, and so on. Thus 1956 equals 0, 1960 equals 5, 1964 equals 3, 1968 equals 1, etc.

Let us point out that the process we have been describing can be applied to the whole Gregorian calendar from its institution (October 15, 1582) to the present time, thanks to a modification introduced by the conjuror Léon Monthenolle. This, as reported by Dr. Dhotel in his masterly work: *La Prestidigitation sans bagages,* is as follows:

"It is first necessary to add to the third table the figure for the leap years from 1840 to 1860, that is: 1840 equals 3; 1844 equals 1; 1848 equals 6; 1852 equals 4; 1856 equals 2. Note that the sequence of these values is identical with those which follow: 3, 1, 6, 4, 2. The third table then becomes the following, with a slight difference of arrangement:

1840	equals	3				
1844	"	1				
1848	"	6				
1852	"	4				
1856	"	2				
1860	"	0	same	value	for	1900
1864	"	5	"	"	"	1904
1868	"	3	"	"	"	1908
1872	"	1	"	"	"	1912
1876	"	6	"	"	"	1916
1880	"	4	"	"	"	1920
1884	"	2	"	"	"	1924
1888	"	0	"	"	"	1928
1892	"	5	"	"	"	1932
1896	"	3	"	"	"	1936

"Remember that 1940 equals 1.

"For dates earlier than 1840 it is only necessary to add what we shall call the 'century-value' to the final figure of the operations shown earlier. This will be:

"If the date corresponds to the column on the right (that is, if it is included within the 39 first years of the century):

2 after 1800, that is from 1800 to 1839 inclusive; 4 after 1700, that is from 1700 to 1739 inclusive; 6 after 1600, that is from 1600 to 1639 inclusive.

"If the date corresponds to the left-hand column (that is if it is included in the 60 last years of the century): 2 after 1700, that is from 1740 to 1799 inclusive; 4 after 1600, that is from 1640 to 1699 inclusive; 6 after 1582, that is from 1582 to 1599 inclusive.

1st example.—*What day was May 3, 1613?*

We shall have:

The day of the month	3
The value of the month (May) ..	1
The value of the leap year immediately anterior to 1913 (equivalent to 1613), that is to say 1912	1
The difference between 1912 and 1913	1
The 'century-value' (included between 1600 and 1639) ..	6
Total ..	12

"From 12 we subtract 7, which gives 5. Therefore May 3, 1613 was a Friday.

2nd example.—*What day was July 14, 1789?*

We shall have:

The day of the month: 14 — 2 × 7	0
The value of the month (July) ..	6
The value of the leap year immediately anterior to 1889 (which is equivalent to 1789) that is to say 1888	0
The difference between 1888 and 1889	1
The 'century-value' (included between 1740 and 1799) ..	2
Total ..	9

"From 9 we take 7, which gives us 2. Therefore July 14, 1789 was a Tuesday."

An unusual application of the calendar operation is the fixing of a date for which the day of the week and the day of the month are known, but of which one of the four figures for the year has been destroyed or effaced. The problem is solved by a series of trials and two solutions are generally obtained. The definitive choice of one of these can usually be established by historic or other considerations.

The life span problem

Trick calculators usually introduce the life span problem after that of the calendar. This consists in giving instantly the number of years, months and days lived by a person at a specific date. Several methods have been furnished, but the best is that supplied by M. Benoit-Bazille. It comes down essentially to subtracting the date of birth from the day after the date given.[1]

Example: *What is the age on August 5, 1959 of a person born on June 1, 1895?*

The problem is set out thus:

1959	8	6
1895	6	1
0064	2	5

The person has lived 64 years 2 months 5 days.

If the number of months and the number of days of the top line are less than the corresponding numbers in the bottom line, the procedure is as follows:

1. Where the days are concerned, the difference is found between the smaller number and 30 or 31 (according to whether the month has 30 or 31 days) and this difference is added to the higher number. The total gives the number of days, but it is necessary to add 1 (which corresponds to 30 or 31 days) to the number of months in the bottom line.

2. For the months the procedure is the same. The difference is found between 12 and the number of months in the bottom line, which number can have been increased by 1 from the calculation for the days. Next the difference so obtained is added to the higher number. The total gives the number of the months. Finally 1 is added to the number of years in the bottom line.

Two examples will illustrate these different operations.

1st example.—*What is the age on August 5, 1959 of a person born on November 3, 1880?*

[1]The lightning calculators solve a problem which is apparently analogous, but in reality much more complicated; they give very rapidly, as we have seen, the number of hours, minutes and seconds lived by a person at a date and hour determined.

The problem is set out:

1959	8	6
1880	11	3
78	9	3

Detail of the operations:

$6 - 3 = 3$

$12 - 11 = 1; 1 + 8 = 9$

$0 + 1 = 1; 1959 - 1881 = 78$

The person has lived 78 years 9 months 3 days.

2nd example.—*What is the age on August 5, 1959 of a person born on November 12, 1880?*

The problem is set out:

1959	8	6
1880	11	12
78	8	24

Detail of the operations:

$30 - 12 = 18; 18 + 6 = 24$ (November has 30 days)

$11 + 1 = 12; 12 - 12 = 0; 0 + 8 = 8$

$0 + 1 = 1; 1959 - 1881 = 78$

The person has lived 78 years 8 months 24 days.

Memory for numbers

Let us see finally how an exceptional memory for numbers can be simulated.

When calculators like Inaudi or Mlle. Osaka engrave a whole series of numbers on their minds, they do not try to associate them with any particular significance. The figures are retained in the memory as they are, whether as articulated sounds or in the form of visual images. There exists no logical link to help the calculator to remember them.

Mnemonics, or mnemotechnics, is specifically designed to substitute a memory of ideas for a memory of sensations. In the field which concerns us here this technique is adopted to give a special and wholly artificial significance to figures, enabling them to be remembered more easily.

We will not include here a lengthy description of the processes of mnemonics, because their detailed study would carry us too far and can only interest specialists. We give here only the essential principles, which are two in number.

The first consists in translating figures and numbers

which have no meaning into as many words which have meanings and to introduce these words into sentences. When this has been done it is only necessary to fix the sentence in the memory and at the opportune moment give them back their figure significance. The translation of figures into words is carried out in the following manner. A concordance has been established between the ten figures and different consonants of the alphabet, the conventions generally adopted being these: 1 is represented by *t*; 2 by *n*; 3 by *m;* 4 by *r;* 5 by *l;* 6 by *j;* 7 by *o;* 8 by *f;* 9 by *p;* 0 by *z*. When this table of correspondence has been learned thoroughly, the letters are substituted for the figures, and words and sentences are then constructed.

An example will make the technique clear. Suppose that 2,345 has to be memorized. We replace each figure by the appropriate consonant and obtain *n, m, r, l*. This first translation has no meaning; it is as difficult to remember *n, m, r, l* as to remember 2,345. But the process does not end there. The consonants supply only the framework of the actual word; to create the word vowels are added to it. Here, with all the freedom allowed to the mnemotechnician, we can form the word *numeral*.

The second principle of mnemonics consists in associating the words into which the figures have been translated with images known in advance, as the ancients did, or in making use of tables of "reminder words," as is done today.

Although the process of mnemonics appears rather hard going at first sight, experience has shown that it can hold its own against the natural memory of lightning calculators. This is the conclusion reached by Alfred Binet in his comparative studies of Inaudi and Diamandi, on the one hand, and M. Arnould, a very able mnemotechnician, on the other.

"Two very different cases must be distinguished for the mnemotechnician," writes M. Binet. "In the first case it is he who selects the figures to be memorized; in the second case he is given figures and must memorize them as they are given to him. When it is the mnemotechnician who has selected the figures he can acquire a facility so great that the number he can retain, thanks to the resources of his art, is *indefinite*. We emphasize this word because the fact seems scarcely credible, and yet it is very easy to understand. M. Arnould has shown us that if he is allowed to recite figures at will, he can repeat several thousands for hours on end; he could dictate a thousand, ten thousand, a hundred thousand or even a million to a zealous secretary, then *he could repeat them all exactly, in the same order, without forgetting any*. We began the experiment and if we did not carry it to its conclusion it was not through lack of patience, but because we were convinced, quite easily, that nothing could be more simple for him.

"Here is the method employed; the mnemotechnician knows by heart a hundred lines of verse, or even two or three hundred; he translates the consonants of the words into figures; he carries out this translation with a dexterity of mind developed by long practice, the two or three hundred verses which he knows by heart supplying him with two or three thousand figures. When he has ended that series he starts again, having made a mental note to vary it, so that in the second series he will increase each figure by one unit; he could, in the same way, for a third series, increase each figure by two units; then in a fourth series make a still larger addition. Let it be noted that the number of these variations is practically indefinite, since, after having made additions, he can do subtractions with all the figures, then multiplications, and so on. He can get up to a hundred thousand figures without any effort; we can willingly suppose that the million could be attained without serious difficulty. All this is very simple when once the key of mnemonics is known; and this key can be created by anyone for his own use and employed after a little practice. It is almost child's play.

"It is somewhat different when the mnemotechnician receives the numbers to be remembered from a third person. The task he has to carry out in the latter case is the reverse of the previous one: instead of translating words into figures, he translates figures into words and seeks to stick these words in his memory so that he can retranslate them into figures when the time comes."

Let us add, to complete these few observations about mnemonics, that there are a large number of special methods for remembering a series of numbers easily. By using the following method, for instance, one can repeat a sequence of points scored by a dice.

It is arranged that a dice shall be thrown by different spectators, each of whom is asked to remember the score he has made, or, better still, to note it on a sheet of paper. When this has been done you then remember the score made by each spectator.

To do this you learn in advance a list of hats numbered from 1 to 6.

1 cap	3 Homburg	5 fedora
2 beret	4 top hat	6 nightcap

and you mentally picture the spectator wearing the head-dress which corresponds to the score he has thrown; you have only to recall this picture to be able to give the result. This experiment is extremely easy to carry out.

An exercise of the same kind, but more difficult to execute, consists of asking the audience for 20 or 30 words and then repeating them, either in the order in which they have been uttered, or starting from the last and working backwards. In addition, the relative position of this or that word, given at random by the public, can be given.

This experiment is carried out by means of a key of 30 words which has been carefully memorized. Each word begins with a letter of the alphabet which corresponds to its position, except the last four words, which must be remembered directly since there are only 26 letters in the alphabet. These can be selected in such a way that they recall the sound of the number. Thus 27 can be, for instance, *heaven;* 28, *slate;* 29, *spine;* 30, *dirty.*

In principle the words of the list should be quite distinct from each other, so that they can be easily fixed in the memory, and very evocative, so that they easily arouse an association of images.

Here is a list which can be used, but the best list is the one you make for yourself.

1	A	Alligator	11	K	Kettle	21	U	Umbrella
2	B	Book	12	L	Lane	22	V	Vegetable
3	C	Cactus	13	M	Mosquito	23	W	Woman
4	D	Dog	14	N	Nurse	24	X	Xrays
5	E	Ear	15	O	Oak	25	Y	Yellow
6	F	Fence	16	P	Pool	26	Z	Zipper
7	G	Gun	17	Q	Queen	27		Heaven
8	H	Hand	18	R	Rabbit	28		Slate
9	I	Ink	19	S	Skeleton	29		Spine
10	J	Jug	20	T	Tray	30		Dirty.

To use this 30-word key you associate the image of the word given by the spectator with the image of the word in the same position in the list. Thus, if the first word supplied is *man,* you imagine a man eaten by an alligator. If the second word is *bedroom* you see a mental picture of a book in a bedroom, and so on. By recalling the word-association table you will be easily able to repeat a series of words given by the audience. Each word will immediately conjure up in your mind the image with which you have associated it. You will even be able to give the position of this or that word supplied by the audience. The word will remind you of the word in the table and therefore of its number on the list.

CHAPTER III

Trick Mathematics and Memorizing Numbers

IN THIS CHAPTER we shall examine some exercises in arithmetic and mnemonics which depend almost wholly upon conjuring and not, as with the preceding exercises, upon the use of more-or-less complicated rules having a mathematical basis. These tricks often form part of a conjuror's repertory and are generally presented as examples of telepathy, precognition or fabulous memory. We will give first the "effect," then the explanation.

The five first tricks and the last two, all of which are very simple, could easily be carried out by the reader.

Mental calculation and "telepathy"

Effect.—On the stage, facing the public, is a blackboard. The calculator is blindfolded and stands to the left and a little behind the board. His partner asks the audience for numbers at random with which various additions, subtractions, multiplications, divisions, etc., are carried out. While he is writing down the results, the calculator repeats them aloud as if he could see them in reality. The performance ends with the partner crossing out the figures one set at a time while the calculator simultaneously names them.

Explanation.—The partner holds the blackboard in a natural way with his left hand, but by means of the fingers of this hand, extended or bent to a greater or less degree, he supplies the calculator with the indispensable information invisibly to the spectators because his fingers are hidden by the blackboard. Some performers do this trick with astonishing skill and the pseudo-calculators announce the right figures with prodigious rapidity.

Needless to say the bandages with which the calculator

is blindfolded is faked. Numerous precautions, which it would be tedious to explain here in detail, enable him to see through it. One of the simplest methods is to pull a few threads out of a thick bandage and place the thin area in front of the subject's eyes. To show that the bandage is "really" opaque the partner tries it over the eyes of a spectator in such a way that the transparent section is over his forehead.

Variation of the trick

This is based on quite a different principle.

You ask a spectator to write down or to say rapidly a number formed of three identical figures (you emphasize the word "three"). Experience shows that he will most often write 333. If you have previously written 333 on the back of a slate or a piece of paper, you show this and say that you have predicted (or "mentally imposed") this number.

But if the spectator gives some other number, you ask him to add up the three figures and divide the number by the answer. You then take out from your pocket a paper on which is written the number 37.

Whatever the number of 3 identical figures the result is in fact always the same:

555 divided by 15 gives 37

777 divided by 21 gives 37.

If the number supplied is 333 you pass on to a second spectator and carry out the second part of the trick. You will then have made two successive "predictions."

Predicting a number

Effect.—Ten squares of paper exactly alike are distributed and each spectator is invited to write a number of several figures on his paper. The papers are then folded twice and collected in a borrowed hat.

A spectator shuffles the papers in the hat and another is then asked to take out one of them at random and hold it up between his thumb and forefinger so that it can be seen by the audience. The performer then simulates a great mental effort and announces a number which it is claimed is that written on the paper held by the spectator. When the paper is unfolded it is found that the number announced is indeed the right one.

Explanation.—A square of paper has been prepared on which a number has been written, 2,653 for example. This folded paper is palmed in the right hand, either before or during the performance. The hat borrowed is a soft felt hat, pleated in the middle so that its inside forms a sort of receptacle with two compartments. You hold the hat in your right hand, with the opening upwards, and the compartment nearest the hand is squeezed tightly against the outer edge of the hat. In these conditions the folded papers are received in the open compartment. When a spectator who is to take out one of the papers is approached, the hat is switched from the right hand to the left and in offering the hat the papers received are squeezed out of sight, while the compartment previously closed is opened. It is into this empty compartment that the palmed paper is dropped. This can be done very easily by pretending to grasp the brim of the hat with your right hand. If you do not feel you could manage this skillfully enough, you can simply place the prepared paper in the hat on a pretext of shuffling the papers. The rest of the trick works automatically. You hold the hat fairly high up so that the spectator who takes the paper out does not see the bottom and so will not notice that only one paper is being offered to him.

The "forcing" of the prepared paper can also be carried out by the aid of a metal holder with a clip at each end. The prepared paper is fixed in one of the clips. This end is hidden in your right hand when you pick up the holder, previously on a table. This time no hat is needed. The folded papers are collected on a tray and the one chosen by a spectator is clipped to the other end of the holder. Another spectator is asked to read out the paper, but in going from the first to the second spectator the conjuror switches the holder to his left hand, reversing the ends as he does so; in this way the paper selected is automatically replaced by the prepared paper.

Magical addition

Effect.—The conjuror announces that he is going to give in advance the result of an addition of which all the figures will be supplied by the spectators. He writes a number on a paper which he folds into four and then gives to a spectator, or else he writes the number on a blackboard turned away from the audience. Next a piece of rather stiff paper, measuring about 3 inches by 4, is placed on a piece of cardboard double this size, to form a pad. A number of 3 or 4 figures is written on the paper by a first spectator, then two other numbers, also of 3 or 4 figures, are written one under another under the first number by a second and a third spectator. A line is drawn under the numbers and a fourth spectator does the addition.

The total is then found to be the number which had been forecast by the conjuror and which had been written on the folded paper or on the blackboard.

Explanation.—On the back of the stiff paper the performer has previously written 3 numbers of 3 or 4 figures of which he knows the sum. While approaching the fourth spectator, who should be some distance away from the first three, he reverses the paper, masked by the cardboard backing. This operation is made possible by the fact that the spectator is a little way away and prevents any check by the spectators who supplied the numbers.

Instead of a piece of paper with a cardboard backing you can use a note pad with identical top and bottom sheets. In this way the substitution is made easily by turning the pad round when it is taken to the spectator who does the addition.

Another version consists in asking two or three people in succession to think of any number of three or four figures. The performer claims that he has read their thoughts and to prove this he writes a number on a blackboard which the public cannot see, claiming that this is the total of the numbers thought of.

He then approaches these spectators and asks them to write the numbers they had thought of on a writing pad which he hands to them.

Finally a last spectator who has not yet taken part is asked to add up the numbers. While he is doing the calculation he writes down the figures very clearly on a piece of special paper which he retains.

But, in approaching this spectator, the conjuror, as before, has surreptitiously reversed the pad so that the sum obtained is the total of numbers which he had previously selected. This is the total which appears on the blackboard.

Before revealing the totals, you can make the trick more convincing by drawing the attention of the first spectators to the fact that it is their numbers which are written on the pad. Of course this is turned round before the verification is made. It is wise to have the numbers crossed out as they are recognized so that no suspicious spectator may have a chance to add them up in his head.

Variation of the trick.—You have a number of 4 figures written down and as soon as this has been done you ostentatiously write a number on a piece of paper which will be shown only at the end of the experiment and which as before will be the result of an addition of which none of the terms, except the first, has yet been given.

You then apply to a second person and ask him to whisper in your ear a random number which you then write underneath the first one. Next you obtain a third number from a spectator some distance away from the first two but this time you only pretend to write it down; in reality you write the number needed to make the second number up to 9,999. To prevent the third spectator from noticing the subterfuge, you do the actual writing while you are going from him towards the two first spectators and you then ask one of these to do the addition. These recognize the numbers they have dictated and do not suspect that the third number has been faked. As for the total you have promised, you obtain this by putting 1 to the left of the first number and taking away 1 from the right.

Example.—You give 2,543 as the first number. At once you write down the eventual sum: 12,542. If the second number is, for instance, 1,432, you write 9,999 less this number as the third number, i.e., 8,567. You will then have the following addition:

```
 2,543
 1,432
 8,567
------
12,542
```

Another variation.—The performer borrows a business card from one of the spectators, and taking a pencil, pretends to write on the back of it a number which, he says, will be the result whatever operation is to be carried out by the audience, addition, subtraction, division or root extraction.

Next, with his right hand, he holds the card vertically, the printed side towards the audience, and keeps it thus while the spectators do the operation of which the nature and the figures have been entirely selected by them.

When this is finished, the operator asks for the answer,

in order, so he says, to check it with what he has written. He then glances at the back of the card and declares that he has predicted the result of the operation, as the spectators can verify immediately.

To do this trick the performer makes use of a small apparatus, absolutely invisible, consisting of a tiny metal blade in the shape of a crescent, fitted with a fragment of very soft graphite and stuck inside the nail of his right thumb. As soon as the number is announced the performer writes it on the back of the card, the movement of his thumb being concealed by the card and by his fingers. Naturally it is best that the number should be short; a division or the extraction of a root is to be preferred.

The calculating slates

Effect.—A predicted number or the result of an addition can be written "magically" on a slate by "spirits." The effect is the following: two slates, preferably imitation (of painted cardboard) are shown, framed in wood and completely blank. These are placed one on top of the other and tied with ribbon. After the evocation of a spirit or other suitable occult entity, it is found, when the slates are separated, that one of them bears the number which has been called out, or the total of the addition which has been performed.

Explanation.—Several methods can be employed to obtain this result. You write the required number on one of the slates beforehand and cover it with a separate sheet of cardboard or metal, of the same color as the slate and exactly fitting into the frame. The slates thus appear to be blank. The rest of the trick needs little explanation. You place the genuine slate on the prepared slate; you tie them with a ribbon or cord; in the course of doing this you reverse the slates so that the lower slate becomes the top slate and vice versa. The separate sheet is now attached to the blank slate and when the two are separated it is seen that one of them bears the desired number, which, needless to say, has been 'forced" by one of the two techniques described before.

Here is an original way of making the figures appear. You rub one side of one of the slates with chalk so that it becomes white. Then, with a wet finger, you trace out the figures which thus come out in black on a white background. You cover it with the separate sheet as before. After having demonstrated that the slates have not been prepared in any way, you smear the visible side of the sheet with chalk and the operation then proceeds as before. You explain that the spirit has produced the writing with its "finger," which is on the whole more plausible than when it is suggested that an ordinary piece of chalk or a slate pencil has been used.

You can also use slates covered with a rectangle of black silk which you whisk out of sight at the desired moment, or, better still, which rolls up into the frame with a spring, but these props cannot be employed at close range.

More subtle methods can be used. For one of them the slates are quite ordinary ones. You have the clean slate examined; then you take the slate bearing the inscription but by a substitution analogous to "palming" you in fact have the first slate examined again.[1]

Two cardboard slates held together by a ribbon sealed with wax can be employed. By pressing the edges you make the two slates bend outwards and you write on the inside face of one of them with a finger fitted with a fragment of pencil lead fixed to an invisible thimble.

In yet another system the slate is wrapped in white glazed paper on which the figures to be made to appear have been written in chalk, in reversed characters. On the pretext of wiping the slate you slightly moisten it, wrap it in the paper, and the inscription is transferred on pressure. You can also trace an invisible inscription with a lithographic pencil. You make this become visible by means of chalk powder hidden in the wiping rag. A similar process consists in painting a slate with a solution of gum arabic and then writing with a slate pencil. The presentation is as follows: you wipe the unprepared side with a sponge dipped in alcohol, then the prepared face. You place the slate on the table and in a little while the writing gradually appears before the spectators' eyes.

Trick predictions

Effect.—When the curtain rises (for this is often done in a auditorium or a theater) the spectators see an envelope

[1] This form of substitution is common in conjuring. It consists essentially in replacing one object, a card, a slate, or a ring, held in one hand and shown, by another object of the same kind, held in the other hand.

sealed with five large wax seals, suspended from the ceiling of the stage. After appropriate patter about telepathy and premonition the conjuror asserts that he is able to foresee the most unlikely events and that he is going to prove it.

"Suspended from the ceiling of this stage," he says, "is a sealed envelope. Inside it is a sheet of paper on which some words have been written. Also in the envelope is the result of an addition which is going to be done freely by the audience."

Next he asks various spectators to write some numbers on a writing pad, one under another, and has these added up by a last spectator who notes the total but does not announce it. The performer now displays a box of dominoes arranged in four rows of seven, making twenty-eight dominoes in all, and asks one person in the audience to select one of the rows. Let us say that the choice falls on the third. Next a card is taken from a pack by a spectator; finally a photograph is chosen from a set which includes some portraits of well-known personalities. The card and the photograph are retained by the spectators who do not reveal what is on them.

This done, the letter is opened by someone from the audience who reads the contents aloud.

"The result of the addition will be 14,526; the row of dominoes will form a total of forty-two points; the card and the photograph chosen will be the queen of hearts and Marilyn Monroe."

All these predictions are then found to be correct.

Explanation.—First of all the total of the addition is known in advance by means of the procedure described earlier under *Magical Addition.* It is known that the sum of the points marked on a set of dominoes is 168, of which the quarter is 42. Four rows each totalling 42 have therefore been arranged, thus:

1st row		*2nd row*		*3rd row*		*4th row*	
6/6	12	6/5	11	6/4	10	6/3	 9
5/5	10	5/4	 9	5/3	 8	5/2	 7
4/4	 8	4/3	 7	4/2	 6	4/1	 5
3/3	 6	3/2	 5	3/1	 4	3/0	 3
2/2	 4	2/1	 3	2/0	 2	0/4	 4
1/1	 2	1/0	 1	0/5	 5	1/5	 6
0/0	 0	0/6	 6	1/6	 7	2/6	 8
Totals	42		42		42		42

The audience can thus nominate any row it likes without endangering the success of the trick. It is wise to point out that the different rows are not identical as it might be supposed that the box contains four rows of the same dominoes and not a complete set.

The card and the photograph are "forced" by one of the several methods used in conjuring of which the simplest is the following: the pack of cards is made up of thirty queens of hearts, only the last cards being different; these are fanned out, not too obviously, before a spectator is invited to make his choice, but when he holds out his hand you offer him the other end of the pack, which is entirely made up of queens of hearts.

For the photograph to be chosen you work in the same way; on looking through the collection of portraits you pass them one at a time from one hand to the other, each time announcing a different name, though in fact all of them are of Marilyn Monroe. As with the pack of cards, two or three extra photographs can, if you prefer, represent different personalities.

Variation of the trick.—A variation of this trick can be wholly concerned with arithmetical calculations. The envelope is again suspended rather high up in the middle of the stage and the conjuror has the spectators carry out a variety of calculations: additions, subtractions, multiplications, extractions of roots, raising to powers, etc., the numbers being selected by the audience in all cases. As the operations are done the results are written up on a blackboard and seen by the audience.

When they are finished the operator climbs on a chair brought by an assistant, unhooks the envelope and gives it at once to a spectator, who reads the contents aloud. The numerical results found in it are exactly those which have been given by the spectators.

The principle of this trick is quite different from the last. When the results of the operations are given successively by the spectators, a confederate hidden in the wings collects them and writes them down on a piece of paper. This is then placed in an envelope identical with the one on the stage and the whole is next placed in a sort of pouch fastened to the back of a chair. It is this chair which is brought in by the conjuror's assistant at the right moment. The conjuror climbs on the chair, as we have said, takes hold of the suspended envelope with the right hand, and, in getting down again, grasps the back of the chair with the same hand, as it is natural for him to do. But at that moment he slips the envelope he is carrying into the pouch and takes out the envelope prepared by his confederate, which contains the answers to the sums. The substitution is done practically instantaneously and is absolutely invisible. The rest of the performance takes place as described.

The trick can be made still more striking by getting two or three spectators to check the piece of paper on which the answers have been written. The method then is as follows. The spectators put their signatures on the piece of paper and immediately afterwards the performer pretends to write on it, with an empty fountain pen, a series of numbers which, he says, are the results of the sums done by the spectators. The paper is folded in four, placed in an envelope and this is seemingly suspended above the stage. In reality when the operator climbs on the chair brought by his assistant he exchanges his envelope for a similar one placed in the pouch beforehand, so that the paper bearing the signatures finally comes into the hands of the assistant who, in the wings, can write the answers on it. After that the experiment continues as before.

The hand of Ibycus

Effect.—A wooden rod is first screwed to the floor of the hall among the spectators; next, a small rectangular frame is fastened to the rod by one of its sides and a sheet of glass is placed above it. Rod, frame and glass have been previously examined by the audience; nothing abnormal is found. A hand made of cardboard or plaster, the "hand of ibycus," [1] which has also been examined by the audience, is placed on the glass sheet. The conjuror invokes the aid of "spirits" or of some other mysterious power, but nothing happens. He persists, and just when the spectators are beginning to doubt the powers of the "medium," the hand is seen to rise and tap the glass. The effect is very striking, all the more because it can be

[1] Ibycus was a Greek poet of the sixth century B.C. He was murdered in a forest by brigands; before dying he claimed as witnesses of the crime the cranes which were flying overhead, by *lifting his hand* towards them. Some time afterwards one of the murderers was present at the Olympic Games and, seeing a flock of cranes passing overhead, cried out imprudently: "There are the witnesses of Ibycus." This resulted in the discovery of the criminals.

easily seen, by looking at it from close range, that the hand is not moved by a wire or thread.

But there is more to come: the hand of Ibycus now shows itself capable of doing arithmetic.

Two dice are handed to a spectator for him to throw on a tray, but the hand announces in advance, by taps, the score which the dice will produce: 8 for example. And, indeed, the upper faces of the two dice thrown on the tray do total 8.

Squares of paper are next distributed in the hall and the spectators are invited to write on them any number of three or four figures. The papers, folded into fours, are collected in a bag attached to a long handle. A spectator takes a piece of paper from the bag at random and holds it, still folded, between his finger and thumb; he holds his hand up so that the audience can see the paper. A second, third and fourth spectator do the same. The hand taps out a number. The papers are unfolded. The total of the numbers written on them is the number announced.

Explanation.—It can be revealed at once that the hand is moved by electricity. Electromagnets are arranged inside the wooden frame and are connected, via the vertical rod, with conductor wires running under the floor of the hall. These wires lead into the wings and are supplied with a switch. The current is provided by batteries or accumulators. A steel bar is hidden inside the artificial hand.

An assistant hidden in the wings applies the current by pressing the switch. The steel bar is attracted by the electromagnets and the hand swings and taps.

The movements of the hand explained, let us now look at the way in which it "foresees" the scores of the two dice and the total of the numbers written by the spectators.

The dice are not as genuine as they seem; they are loaded. These conjuror's accessories are dice which have been weighted with lead at one of their faces and whose angles have been slightly rounded. When thrown they roll rather than tumble and always stop in the same position, which corresponds, on the upper face, to a particular score. The figure 8 would be furnished for instance by two dice giving 5 and 3 respectively. Naturally the conjuror will have several dice available, ballasted in different ways, if he intends to repeat the trick.

The assistant in the wings will then press the switch as many times as are necessary, whatever the number.

The prediction of the addition total depends upon a

principle analogous to that used in the trick with the "predicted number."

The collecting bag has two pockets, either of which can be opened by means of a steel rod running through the handle and ending at its extremity. Four squares of paper, carefully folded, and each carrying a different number, the whole forming a total known to the performer, are placed beforehand in one of the compartments, in the right-hand compartment for example. This compartment is closed and the bag appears to be empty. The papers handed to the spectators are identical in size and color to those which are in the bag; these are collected in the left-hand pocket. When the operation is completed the rod is set in action and the left-hand pocket closes, leaving the right-hand one open. In fumbling in the bag the spectator necessarily draws the four papers bearing the numbers of which the total is known to the performer and his assistant. The latter makes the hand announce the answer in the same way as for the dice. A fresh manipulation of the bag mechanism closes the right-hand pocket and opens the left. Investigation by spectators is now prevented. The genuine papers are dropped on the stage and the performer can if necessary have some of them identified, but this is not necessary.

Various improvements of this trick have been suggested. These include the use of compressed air, faked trays and the use of wires controlled by an assistant or by the performer himself. In this last case, which is handy for the amateur, the conjuror has a black thread attached to a ring on the left middle finger. He holds in his two hands a tray on which stands the hand of Ibycus. In making the appropriate gestures to demonstrate that there are no threads, either above or below the tray, he puts in place the thread which was hanging down the back of his left hand. From then on the thread runs over the wrist of the artificial hand, while its free end is held by the conjuror's right thumb. A slight movement of the left little finger, concealed under the tray, pulls the thread and moves the hand. The device is absolutely invisible, even from close range.

The invisible brain

Effect.—The performer goes on the stage holding in

his right hand a skull with no top, which can appear to be genuine or can be artificial, i.e., of painted cardboard.

"This skull," he says, addressing the spectators, "once belonged to a lightning calculator, and wonderful to relate, it is capable of performing the mathematical operations which the calculator carried out during his lifetime."

He then puts the skull, with its opening upwards, on a small table standing against the backdrop, this being made of a cloth with vertical stripes, and distributes some squares of paper to a few spectators. The papers, which are completely blank, are folded in four and one of them is handed to the performer; he puts it in the skull. A first experiment is made; for instance the one described under the heading *Magic Number*. A spectator is then invited on to the stage to take the paper from the skull and unfold it; the answer to the sum will be found written on it. The experiments continue and the "invisible brain" which is supposed to be inside the skull gives the answers to various arithmetical operations, predicts the total of an addition, the score produced by dice thrown by a spectator, etc.

Explanation.—The skull is placed on the table, which is fairly high, at the level of a slot cut in the back cloth, along one of the vertical stripes. It is thus invisible. An assistant hidden behind the back cloth carries out each operation and transcribes the result on a square of paper which he folds and pushes through the slot and into the skull with the aid of a small pair of pincers. Naturally the spectators' squares of paper are successively disposed of, so that nothing is, in fact, put into the skull by the performer.

The telephone directory and the newspaper

Effect.—The performer claims that he has learned the contents of a telephone directory by heart and that he is able to give the text of any column or line as soon as it is nominated. An assistant then places several pads of cotton wool round his head, tied in place by bandages to leave him no possibility of seeing. Thus blindfolded, he is led to the center of the auditorium where he remains, either seated or standing. Then the assistant hands him a directory which has been examined by a number of spectators, chosen at random. The directory is not faked.

The spectators open it at any page and ask the conjuror to give the text of a specific line in a specific column on a specific page. To the great astonishment of the audience the illusionist immediately does so. The trick is repeated several times at various points in the auditorium with the same success. Finally to demonstrate yet more strongly that there is no faking the assistant stands the illusionist on a chair covered with a thick glass plate, thus isolating him from the floor. The result is equally successful.

Sometimes the directory is replaced by a current newspaper and, as in the previous experiment, the conjuror gives the required text.

Explanation.—To perform one or other of these tricks, use is made either of an ordinary telephone, or, preferably, a miniature short wave radio set. In the first case a telephone is installed in the wings. The wires leaving it pass under the auditorium carpet and end at the point where the performer stands, their ends forming a slight ridge. The so-called memory expert wears shoes with soles made of copper sheeting. From these sheets run wires which go up his legs and body and end under the collar of his jacket. Tiny telephone receivers are concealed in the cotton wool pads. The pressure of the bandages which are supposed to cut off his sight puts these wires in contact with the receivers and ensures communication with the telephone.

When the illusionist is led among the spectators he closes the circuit by putting his feet on the ends of the wires coming through the carpet and is thus put in contact with an assistant in the wings. As soon as the latter hears the spectator ask for the text of a selected line, he opens the required page of an identical directory and telephones the text to the illusionist.

In order to put off suspicions which could arise if the trick were always done at the same point, there are several electric contacts so that he can change his position. Even the chair, despite appearances, does not isolate the subject from the floor. The telephone wires, passing through the carpet, come into contact with other wires cleverly concealed in the chair legs. The chair wires emerge near the back of the seat, and are bent, at the required moment, over the glass sheet, to make contact with the copper shoe soles.

Naturally all this somewhat complicated installation is rendered unnecessary by the use of a small short-wave radio. The assistant has the transmitter, often in the form of a wrist watch and the receiver is concealed under the performer's clothes or hair. An apparatus of this type has been on sale in the United States since 1937. It is frequently used in demonstrations of so-called "thought transference" on the stage.

Variation of the trick.—On the stage is a shelf of books including various directories. The memory expert, whose eyes are bound, is seated on a chair, facing the audience. In front of him is a microphone. His partner stands beside him. Several spectators are invited to go on the stage. One after another they choose lines from the books, and as soon as the choice has been made, the performer announces (through the microphone) the words in the line nominated.

As before, an assistant in the wings has the same books before him as those on the stage. As soon as the title, page and line number chosen are announced, he looks up the required text; but here it is *he himself, who transmits the text to the audience by means of a second microphone.* The memory man merely makes the motions of speaking to the audience through the microphone.

This trick can easily be adapted as an exhibition of "telepathy." In this case without a word being said the spectator indicates the line chosen with his forefinger. The memory man's partner conveys this information to the assistant by means of signs. Only three figures are transmitted: corresponding respectively to a book, a page and a line.

Predicting the end of a game of dominoes

Effect.—A box of dominoes is emptied out on a table, points side down, and they are then shuffled. Two or three people are then asked to play a game taking dominoes at random, as is done normally, and not "cancelling." As soon as each player has put down one or two dominoes it is announced on what "spots" the game will end. And in fact it does so end. The result appears astonishing, especially when it is realized that the total of combinations which can be obtained with 28 dominoes can be expressed by the colossal number:

10,752,728,122,249,860,612,096,000.

Explanation.—In mixing up the dominoes you palm one

of them in the left hand and surreptitiously note it when the game has started and the players are occupied. The game will end on the same points as those found on this domino, on condition that it is not a "double" and that there is no cancelling. These then are the spots you announce. At the end of the game the domino is slipped back into the set while you are helping to turn them over. The set is now once again complete and the trick can be done again.

Variation of the trick.—The dominoes are set out, according to the rules of the game, while you have your back turned. You then ask for one of the dominoes and as soon as you have made a note of it you state what the end spots of the chain will be. In reality this domino does not enter into the matter; it is asked for solely in order to put the spectators off the scent. As in the preceding trick, you have in fact secretly obtained possession of another domino while the set was being scattered and it is this domino which supplies the end spots.

CHAPTER IV

Some Unusual Problems

DURING THEIR performances illusionist calculators often set their audiences some odd or unusual problems, to which the answer is generally an unexpected one. As there is a large number of these, this chapter would be very long if an attempt were made to include them all, and in so doing we should stray too far from our main theme.

This chapter has been added principally because I have noticed that schoolboys and students, of whom there will perhaps be some among my readers, are often attracted or amused by this type of question. Moreover it is worth while to know something about them, in order not to appear ignorant if they arise. One may even wish to introduce them oneself on occasion.

Progressions. The problem of the grains of corn

Arithmetical and geometrical progressions, particularly the latter, result in numbers which can be extraordinarily large and for this reason are unforeseeable *a priori*. The best known and oldest of these problems is that of the grains of corn. According to legend, when Sessa had invented the game of chess, the king of Persia asked him what he wanted as a reward. He was ready, he said, to offer part of his kingdom. With a modest air the artful mathematician asked for a grain of corn for the first of the 64 squares of his chessboard, 2 grains for the second, 4 for the third, 8 for the fourth and so on, doubling the number in each case up to the last. The courtiers thought him stupid to ask so little, when he had been offered riches. But Sessa showed that the king could not satisfy his request, however rich he was. In fact the grains of corn reckoned according to the number of squares on the chessboard form the 64 terms of a geometrical progression having 1 for its first term and 2 for its ratio. Their sum S is given by the well known formula:

$$S = a \times \frac{q^n - 1}{q - 1}$$

in which we have: n = 64; q = 2; a = 1.

Replacing the letters by their values we obtain:

$$S = 1 \times \frac{2^{64} - 1}{2 - 1} = 2^{64} - 1$$
$$= 18{,}446{,}744{,}073{,}709{,}551{,}615.$$

To make this 20 figure number to some extent tangible, let us add that the total production of the earth, throughout several centuries would not be enough to produce such a number of grains of corn. Furthermore, to count them, at the rate of 5 per second, nearly 1,170 million centuries would be needed.

The $24 of the Indian Chief

This problem is equally well known under a variety of titles, but is nonetheless worth recalling. It is usually set out like this: if the Indian chief who sold Manhattan Island for $24 in 1626 had invested his money at 4 per cent, what would his estate be worth now?

Solution.—Compound interest means, of course, that at the end of each unit of time the accrued interest is added to the capital, so that the capital and accruing interest on it are constantly growing. The usual unit of time is one year. So problems dealing with compound interest generally run as follows: *Calculate the final amount of a capital A invested at compound interest for* n *years, at the rate* r, *the accuring interest being added to the capital at the end of each year.*

Now, it can easily be shown that the value acquired (which is A^1) by the capital at the end of n years is:

$$A^1 = A\,(1 + r)^n$$

This formula produces calculations which are easy to do by logarithms, for you have:

$$\log A^1 = \log A + n \log\,(1 + r).$$

If this method is applied to the particular problem of the Indian chief's $24, an incredible result is arrived at. In fact, a capital is obtained which would enable the chief to buy the Chrysler Building, the world's second tallest skyscraper.

It is worth noting that if the $24 had been invested at 6 per cent, the chief's descendants could be rich enough to buy Rockefeller Center sixty times over, or the Empire State Building 100 times over. On the other hand, if it had been invested at a rate of 2 per cent, the $24 would yield a capital of less than $20,000, just about enough to buy a small, modest home. So you see how important interest rates can be.

The Problem of the Four Horseshoes

The problem of the four horseshoes, which is also based on progressions, involves a number less "astronomical" than the last but is nevertheless impressive. Here are the terms: a blacksmith proposes to fix a price of 1 penny for the first nail, two cents for the second, four cents for the third, eight cents for the fourth and so on up to the 32nd nail, each shoe requiring 8 nails. What is the total cost of the shoeing?

Solution.—The following formula is applied:

$$S = a \times \frac{q^n - 1}{q - 1}$$

in which we have: $n = 32$; $q = 2$; $a = 1$, and we obtain:

4,294,967,295 pennies.

The total cost of shoeing will therefore be $42,949,-672.95.

The Problem of the Folded Paper

This is an amusing problem, which you can present as follows:

"Ladies and Gentlemen," you say in as casual a tone as possible, as you begin negligently folding a piece of thin paper, "you see that what I am doing is not very complicated; I am simply folding this quite ordinary piece of paper. Then I fold it a second time" (this is done at right angles to the first fold) "then again, and so on. You can see that this is neither lengthy nor difficult" (here you stop at the fourth folding, while still continuing to talk) "it takes no more than a second to make each fold, does it? If I did it, say fifty times in succession" (while talking you have gradually unfolded the paper) "it would take no more than a minute.

"Now, suppose that I did do this fifty times—not that I wish you to waste a whole minute in watching me do it —about how thick do you think the little packet of paper would be, supposing that the paper itself is a tenth of a millimeter thick? Naturally, I am not taking into account the air in between if the paper is not pressed tight down."

Many people, thinking of fifty thicknesses of paper, will say: "five millimeters." Others, scenting a trap, will increase the figure and will say, completely at random, "several yards." Some, perceiving that a question of geometrical progression appears to be involved, will give numbers up to several miles. But it is probable that no one will come anywhere near the real figure.

"Well now, ladies and gentlemen, the little packet of paper fabricated before your eyes in less than one minute will have a thickness of *just over 70,368,681 miles,* or more than two-thirds of the distance from the earth to the sun!"

This is, in fact, 2 to the power of 50.

Be careful not to say: "I fold it in two, then in four, then in eight . . ." These numbers, increasing rapidly, would give a clue towards the size of the final figure. Say: "I fold it once, then a second time, now a third . . ." and insist on the short time needed to do fifty folds.[1]

The Problem of the Invitations

Eight people plan to invite one another mutually to dinner until they have exhausted all the ways in which they can be seated at table in relation to each other.

Solution.—The possible number of combinations is given by the product:

$$1 \times 2 \times 3 \times 4 \times 5 \times 6 \times 7 \times 8 = 40{,}320.$$

The eight people must therefore dine together for 40,320

[1] The phenomena of cell multiplication lead to results of this kind. Thus a bacterium can produce a *trillion trillion trillions* of descendants in four days; and it has been calculated that infusoria (about 1/10 of a millimeter in diameter) dividing at a rate of four or five times a day would produce in one month a volume of progeniture, provided food and space were available, equal to *one million times the volume of the sun.*

days, or for 110 years 4 months 22 days (leap years being taken into account). In consequence they certainly would never be able to carry their plan into practice.

An even larger number of guests can, of course, be considered. Thus with 13 people the maximum number of combinations is:

$$1 \times 2 \times 3 \times 4 \times 5 \times 6 \times 7 \times 8 \times 9 \times 10 \times 11 \times 12 \times 13 = 6{,}227{,}020{,}800.$$

If the dinners take place only a hundred times a year it would be necessary to wait seventy-two million, two hundred and seventy thousand, two hundred and eight years for all the combinations to be exhausted.

The Ignorant Mathematician

Under this title Mr. R. V. Heath published an interesting arithmetical curiosity in the magical review *Sphinx*.

"Recently," he says, "I met an ignorant mathematician, who told me he was incapable of multiplying or dividing other than by the number 2, and yet he claimed to be able to multiply any number of two figures rapidly.

" 'Right,' I said to him, 'what is 75 times 38?'

"He at once took a sheet of paper and wrote the number 75 on the left and the number 38 on the right. Then he continued as follows:

" 'Half of 75 is 37, isn't it?'

" 'No, it is 37.5,' I replied.

" 'Yes, but that is a fraction and I don't know fractions . . . or anyway I have forgotten them . . .'

"And he wrote down 37. Next he went on the divide 37 by 2 and made the answer 18; disregarding my objection he wrote 18 under the 37 and, dividing again by 2, he got 9, which he also divided by 2, giving 4, which became 2 and finally 1 . . . He next did the same with the number 38 on the right-hand side of the paper, but this time he multiplied 38 by 2, which gave 76, which, again multiplied by 2, produced 152, and each time he multiplied by 2, obtaining successively 304, 608, 1,216 and finally 2,432 . . . In other words he multiplied 38 by 2 and then each number by 2 until he had as many numbers in the right-hand column as in the left; at this moment the operations produced the following little table:

75	38
37	76
18	152
9	304
4	608
2	1,216
1	2,432

"Next he informed me that he never bothered with the even numbers in the left-hand column nor the corresponding numbers in the column on the right; he crossed these out with a pencil, so that there remained:

75	38
37	76
9	304
1	2,432

"Finally he added the numbers in the right-hand column, that is, those he had not crossed out, and he obtained:

$$38 + 76 + 304 + 2{,}432 = 2{,}850.$$

" 'And there's the answer: 2,850,' " he told me.

"Immediately I set to work on the multiplication of 75 by 38 in the usual way. The answer was indeed 2,850. Need I say that I was amazed? He took care to add that he could use his system with any numbers he liked."

The String Round the Equator

A string exactly encircles the Equator. It is cut at one point and a piece of string one yard long is added to its length. How far out from the Equator can it now be placed?

Solution.—The usual answer is that this distance is extremely small and that it is of the order of a millionth of a millimeter or even less.

In reality calculation shows that it is equal to about *5½ inches.*

In fact, if R is the earth's radius, the length C of the string circling the Equator is:

$$C = 2\pi R \quad (1)$$

If we add 1 yard to the circumference and let x be the distance of the string from the ground we have:

$$C + 1 = 2\pi(R + x) \quad (2)$$

Subtracting (1) from (2) we have:

$$1 = 2\pi x.$$

Hence:

$$x = \frac{1}{2\pi} = 5.7 \text{ inches.}$$

It is the great size of the earth's radius which leads *a priori* to an erroneous solution. But, as can be realized, the radius does not enter into it, so that the answer is always the same, whatever the dimensions of the sphere.

The problem can also be set in this way: a string encircles the Equator at a distance of 1 yard from the ground. It is shortened by 1 yard and the two ends are then joined. How much nearer is it to the ground?

The Problem of the Apples

Three people are selling apples at market; they have 20, 30 and 40 apples each respectively; they all sell them at the same price per apple and yet bring back the same sum. How does this come about?

Solution.—The sale takes place in two stages: the first time the apples are sold at 1 penny apiece.

The first person sells 2 and receives 2 cents;

The second person sells 17 and receives 17 cents;

The third person sells 32 and receives 32 cents.

The second sale takes place at the rate of 3 cents per apple.

The first person sells the 18 apples which he has left for 54 cents;

The second sells the 13 apples he has left for 39 cents;

The third sells the 8 apples he has left for 24 cents.

In this way they all bring in the same sum, namely 56 cents.

It will be seen immediately that we can vary the price of the apples as much as we like, as soon as we note the one-third relationship of the first sale to the second. We can also vary the numbers of the apples by doubling, trebling, etc., the original number.

This problem, in appearance relatively simple, is in reality very complicated. Treated by algebra it leads to a system of two equations with eight unknowns of which the solution and development require at least twelve pages of calculation. It is a problem of indeterminate analysis.

Here are the equations:

Let it be supposed that three people bring respectively A, B and C apples.

1*st sale:* the apple is sold at a cents.

The first person sells x apples and receives ax cents.

The second sells y apples and receives ay cents.

The third sells z apples and receives az cents.

2*nd sale:* the apple is sold at b cents.

The first person sells $A - x$ apples and receives $b\ (A - x)$ cents;

The second sells $B - y$ apples and receives $b\ (B - y)$ cents;

The third sells $C - z$ apples and receives $b\ (C - z)$ cents.

Totals:

The first person has received $ax + b\ (A - x)$ cents.
The second has received $ay + b\ (B - y)$ cents.
The third has received $az + b\ (C - z)$ cents.

And, by hypothesis, we have:

$$ax + b\ (A - x) = ay + b\ (B - y) = az + b\ (C - z)$$

which constitutes, as we have said, a system of 2 equations with 8 unknowns, if A, B, C are considered as unknowns.

The Problem of the Two Cars

Two cars are making a return journey between two towns 200 miles apart. The first car does the journey there at a speed of 50 miles an hour and the return journey at a speed of 40 miles an hour. The second car makes its journeys both ways at an average speed of 45 miles an hour. Which of the two cars takes the less time to cover the whole distance?

Solution.—As the speed of the second car is the average between the first and second journeys of the first car, it appears *a priori* that the two cars cover their distance in the same time. This is not so. The second car takes less time than the first to do it.

In its first journey the first car in fact takes $200 : 50 = 4$ hours, and for the return $200 : 40 = 5$ hours, or *9 hours in all.*

The second car, for the whole 400 miles represented by the journey both ways, takes:

$$400 : 45 = 8 \text{ hours } 53 \text{ minutes.}$$

The Fly and the Trains

Two trains travelling in opposite directions on the London—Manchester line will pass each other in 30 minutes. One is travelling at a speed of 60 miles an hour and the other at a speed of 55 miles an hour.

A fly starts from the locomotive of the first train, flying at an average speed of 65 miles an hour; it travels as far as the locomotive of the second train, then returns to the first, naturally covering a shorter and shorter distance each time as it repeats this performance until the moment when the two trains meet. How many miles does the fly cover in all?

Solution.—To solve this problem one is tempted to construct a graph representing the flights of the fly or to introduce the terms of a diminishing progression. The solution is much simpler. The fly flies for 30 minutes, and as its average speed is 65 miles an hour, it has covered 32½ miles.

The Problem of the Two Waterlilies

A tropical waterlily of extremely rapid growth enlarged so much that each day it covered a surface double that which it filled the day before, so that at the end of the thirtieth day it entirely covered the lake in which it grew. How long would it take two waterlilies of the same size at the outset and the same rate of growth to cover the same pool?

Solution.—One is tempted to answer that if it takes 30 days for one waterlily to cover the pool by doubling its surface every day then it would take two waterlilies half this time, or 15 days.

This answer would be wrong. For since the plant doubles its surface every day, it will have half its final surface by the end of the 29th day and will in consequence then cover half the pool. Therefore with two waterlilies the result will be attained at the end of the 29th day and not the 15th day.

A Peculiar Number

The conjuror writes the number 142,857 on the blackboard and demonstrates that it has unusual properties.

He multiplies it by 1, 2, 3, 4, 5, 6 and points out that the products are always expressed by the same figures, which are those of the original number, and that these figures permute in a certain order:

$$142{,}857 \times 1 = 142{,}857$$
$$142{,}857 \times 2 = 285{,}714$$
$$142{,}857 \times 3 = 428{,}571$$
$$142{,}857 \times 4 = 571{,}428$$
$$142{,}857 \times 5 = 714{,}285$$
$$142{,}857 \times 6 = 857{,}142.$$

But after the multiplier 7 the figures apparently tire of their cohesion and separate. With 7 as the multiplier we in fact get 999,999. However it will be noticed that the sum of these six figures is 54, which is double that of the figures of each of the preceding products. Furthermore this "fatigue" is only temporary, for the tired figures ask for nothing better than to re-unite, provided they are "helped" a little.

Let us multiply 142,857 by a number as high as we like, for example 32,284,662,474. The product is: 4,612,-090,027,048,218. In this product there is no longer any trace of the group 142,857, but if we split up the product we have obtained into groups of six figures starting from the right, and then add up these groups, we get this:

$$\begin{array}{r} 048{,}218 \\ 090{,}027 \\ 4{,}612 \\ \hline 142{,}857 \end{array}$$

But sometimes the figures recover their "cohesion" with greater difficulty and the addition has to be done a second time.

Thus by multiplying the peculiar number by 45,013,648 we get 6,430,514,712,336.

On splitting this product into groups of six figures starting from the right and adding them up we get:

$$\begin{array}{r} 712{,}336 \\ 430{,}514 \\ 6 \\ \hline 1{,}142{,}856 \end{array}$$

Again taking a group of six figures starting from the right we have 1 left and we again add this to 142,856.

We are then once more confronted by the peculiar number:

$$\begin{array}{r} 142{,}856 \\ 1 \\ \hline 142{,}857 \end{array}$$

This number has still further curious properties.

We divide it into two groups of three figures: 142 and 857. The second figure of the first group, multiplied by its position number, gives the first figure of the second group:

$$4 \times 2 = 8$$

The sum of the two first figures of the first group gives the second figure of the second group:

$$1 + 4 = 5.$$

The sum of the three figures of the first group gives the third figure of the second group:

$$1 + 4 + 2 = 7.$$

Finally, if we again take the products obtained by multiplying the peculiar number by 1, 2, 3, 4, 5, 6 and arrange them one under the other, in a table, we find that the sum of the figures of each horizontal group and the sum of the figures in each vertical column are all equal to 27.

$$\begin{array}{cccccc} 1 & 4 & 2 & 8 & 5 & 7 = 27 \\ 2 & 8 & 5 & 7 & 1 & 4 = 27 \\ 4 & 2 & 8 & 5 & 7 & 1 = 27 \\ 5 & 7 & 1 & 4 & 2 & 8 = 27 \\ 7 & 1 & 4 & 2 & 8 & 5 = 27 \\ 8 & 5 & 7 & 1 & 4 & 2 = 27 \\ \hline 27 & 27 & 27 & 27 & 27 & 27 \end{array}$$

Other Curious Numbers

Let us consider a fraction which has one as its numerator, and as denominator an uneven multiple of 9, non-divisible by 5: for example one of the following fractions:

$$\frac{1}{9 \times 3} = \frac{1}{27}, \quad \frac{1}{9 \times 7} = \frac{1}{63}, \quad \frac{1}{9 \times 9} = \frac{1}{81}$$

$$\frac{1}{9 \times 11} = \frac{1}{99}, \quad \frac{1}{9 \times 13} = \frac{1}{117}, \quad \frac{1}{9 \times 17} = \frac{1}{153}$$

If we transform it into a decimal number by dividing

the numerator by the denominator, a simple recurring decimal is obtained as the quotient.

Now, the whole number formed by taking one, or several, decimal periods possesses the following remarkable characteristic: multiplied by the ninth of the denominator of the fraction in question, this number gives a product which is formed only with the figure one.

The result is, that multiplied by a double quantity, it will give a product only composed of twos; multiplied by a treble quantity, it will give only threes, etc. Finally, if multiplied by the denominator of the fraction, it will give a product only of nines.

1st example:

$$\frac{1}{27} = 0.037 \ldots\ldots$$

$$\begin{aligned}
37 \times 3 &= 111\\
37 \times 6 &= 222\\
37 \times 9 &= 333\\
37 \times 12 &= 444\\
37 \times 15 &= 555\\
37 \times 18 &= 666\\
37 \times 21 &= 777\\
37 \times 24 &= 888\\
37 \times 27 &= 999
\end{aligned}$$

2nd example:

$$\frac{1}{153} = 0.0065359477124183$$

$$\begin{aligned}
65\,359\,477\,124\,183 \times 17 &= 1\,111\,111\,111\,111\,111\\
65\,359\,477\,124\,183 \times 34 &= 2\,222\,222\,222\,222\,222\\
65\,359\,477\,124\,183 \times 51 &= 3\,333\,333\,333\,333\,333\\
65\,359\,477\,124\,183 \times 68 &= 4\,444\,444\,444\,444\,444\\
65\,359\,477\,124\,183 \times 85 &= 5\,555\,555\,555\,555\,555\\
65\,359\,477\,124\,183 \times 102 &= 6\,666\,666\,666\,666\,666\\
65\,359\,477\,124\,183 \times 119 &= 7\,777\,777\,777\,777\,777\\
65\,359\,477\,124\,183 \times 136 &= 8\,888\,888\,888\,888\,888\\
65\,359\,477\,124\,183 \times 153 &= 9\,999\,999\,999\,999\,999
\end{aligned}$$

The Worm in the Dictionaries

Three dictionaries of 1,000 pages each, 3 inches thick, plus ⅛ inch each side for the binding, stand in a bookshelf. The three volumes together therefore form a total thickness of 9¾ inches. A worm starts at page 1 of the first volume; it bores through the paper and cardboard and reaches page 1,000 of the third volume. What distance has it travelled?

Solution.—The answer most often given is 9½ inches, or 9¾ inches less the thickness of the two bindings, that at the beginning of the first volume and that at the end of the third.

In reality the right answer is 3½ inches, for page 1 of the first volume is not on the left of the book, but on the right, and page 1,000 of the third volume is not on the right of it, but on the left. In consequence to go from page 1 of the first volume to page 1,000 of the third, the worm has only to bore through the cover of the first volume, the whole of the second volume and the cover of the third volume. It will thus have travelled:

⅛ in. + ⅛ in. + 3 ins. + ⅛ in. + ⅛ in. = 3½ ins.

The Climbing Snail

A snail is at the foot of a wall 30 yards high; it climbs up 2 yards during the day, and slides down one yard during the night. How long will it take the snail to reach the top of the wall?

Solution.—The usual answer, but the wrong one, is 30 days, since the snail climbs up one yard every 24 hours.

The right answer is 29 days.

For, at the beginning of the 29th day, the snail will be 28 yards up, and, on that day, it will climb the two yards it had left to reach the top.

The Problem of the Cigarette Ends

A collector of cigarette ends can make one cigarette from four ends; he collects 11. How many cigarettes can he make and how many cigarette ends will be left over?

Solution.—The usual answer, and the wrong one, is generally: 2 cigarettes (made with eight cigarette ends), and three ends left over.

The right answer is three cigarettes, and two ends left over. In fact, with the four first ends, the collector will make one cigarette, but he will have one cigarette end left, which, with three others, will make a second cigarette, after which he will have one end left over, which, with three others, will make yet another cigarette. From this last one he will get one cigarette end, which, with his original eleventh, will make two cigarette ends.

The Arab's Testament

An Arab died, leaving his fortune of 17 camels to his three sons; he left half to the eldest, a third to the second son and a ninth to the youngest. How did the three sons divide up the camels?

Solution.—The three sons tried in vain to carry out the terms of their father's will, but found themselves constantly held up by the impossibility of obtaining a whole number of camels for any of the shares. In despair they appealed to the wisdom of the Cadi. The Cadi came upon the scene riding a camel of his own, which he put with the 17 others, and then said to the eldest son:

"The will gives you the right to a half of the 17 camels; I however will give you the half of 18 camels, or 9 camels."

He said to the second son: "You are due to receive a third of 17 camels; take a third of 18 camels instead, that is 6 camels."

Then to the third: "Your share is a ninth of 17 camels; take a ninth of 18 camels, that is 2 camels." Then the Cadi went away, mounted on his own camel, leaving the three heirs delighted with their good fortune, each of them having obtained a larger share than that provided for in the will. For in fact:

the first had: $$9 - \frac{17}{2} = \tfrac{1}{2} \text{ of a camel extra,}$$

the second had: $$6 - \frac{17}{3} = \tfrac{1}{3} \text{ of a camel extra,}$$

the third had: $$2 - \frac{17}{9} = \tfrac{1}{9} \text{ of a camel extra.}$$

And yet the total number of camels they possessed was: $9 + 6 + 2 = 17$, or in fact what their father had left them.

This curious result, which seems paradoxical at first, becomes clear if we note that the sum of the fractions $\frac{1}{2}$, $\frac{1}{3}$ and $\frac{1}{9} = \frac{17}{18}$ and not a whole number, and that in consequence in following the instructions of their father to the letter and supposing that it had been possible to carry out the division, a part of the legacy, i.e., $\frac{1}{18}$th of it would have been left without an owner.

We can take as the number of animals to be divided a prime number P such that it will be equal to the sum of three divisors of P + 1. Thus we get:

Number of Animals P	*Divisors of P + 1*	*Each share*
11	6, 3, 2	$\frac{1}{2}$, $\frac{1}{4}$, $\frac{1}{6}$
17	9, 6, 2	$\frac{1}{2}$, $\frac{1}{3}$, $\frac{1}{9}$
19	10, 5, 4	$\frac{1}{2}$, $\frac{1}{4}$, $\frac{1}{5}$
23	12, 8, 3	$\frac{1}{2}$, $\frac{1}{3}$, $\frac{1}{8}$

Another Strange Share-out

Two Arabs, one carrying 5 loaves, the other 3 loaves, met a traveller who was wealthy but very hungry. They ate together, then the traveller, as the price of his meal, gave them 8 pieces of gold. How were they to share it out?

Solution.—The first Arab, the one who had had 5 loaves, considered that he should receive 5 pieces of gold, the other 3 going to his comrade. The latter replied that he should get 4 pieces and that he would reimburse the former the price of one loaf. However, they could not come to any agreement.

The Cadi solved the problem thus:

"You are both in the wrong," he said. "What has happened is that you have portioned out each of your loaves into 3 equal parts, which makes a total of 24 parts. Of these you have each eaten 8. He who had 5 loaves or 15 parts parted with 15 — 8 = 7 to the traveller; he who had 3 loaves or 9 parts gave him only 9 — 8 = 1 part. The reward therefore is 7 pieces of gold for the one who had 5 loaves and only one piece for the other."

The Milkmaid's Measure

A milkmaid has 8 pints of milk to sell, contained in an 8-pint vessel. Besides this she has two measuring cans, one holding 5 pints, the other 3 pints. A customer presents herself without a milk container and asks her for 4 pints of milk. How does she give her 4 pints?

Empirical solution.—The milkmaid first fills her 5-pint can, taking the milk from the 8-pint vessel; then she fills her 3-pint can, taking the milk from the 5-pint can; 2 pints of milk remain in the 5-pint can. As the customer has no container, she pours these 2 pints into her own 3-pint can, having first emptied the 3-pint can into the 8-pint vessel; she has thus separated 2 pints into her 3-pint can and 6 pints into her 8-pint vessel.

She now starts again, as before, by filling her 5-pint can from the contents of her 8-pint vessel; then she fills up her 3-pint can from the contents of her 5-pint can. The 4 pints asked for now remain in the latter, and by pouring the contents of her 3-pint can into her 8-pint vessel she will have the other 4 pints in her original container.

The operations she has carried out can be set out in the following table: each horizontal line corresponds to one operation and gives the quantity of milk contained in each vessel, the total of the line always yielding 8 pints of milk.

Operations	*8-pint vessel*	*5-pint can*	*3-pint can*
Original position ..	8	0	0
1st operation ..	3	5	0
2nd " ..	3	2	3
3rd " ..	6	2	0
4th " ..	6	0	2
5th " ..	1	5	2
6th " ..	1	4	3
7th " ..	4	4	0

Algebraical solution.—Let us note first of all that the 8-pint vessel does not enter into the solution, which depends only on filling and emptying the 5- and 3-pint cans. In these conditions let us suppose that the 5-pint can must be filled x times and the 3 pint can y times in order to obtain the result required, namely 4 pints.

We then have the following equation:

$$5x + 3y = 4.$$

This equation with two unknowns admits of whole number solutions only, either positive or negative; its solution shows that these are infinite in number; the two simplest being:

$$x = -1 \qquad y = +3$$
$$x = +2 \qquad y = -2$$

DHG

The first solution is translated by the following table:

Operations		*8-pint vessel*	*5-pint can*	*3-pint can*
Original position	..	8	0	0
1st operation	..	5	0	3
2nd "	..	5	3	0
3rd "	..	2	3	3
4th "	..	2	5	1
5th "	..	7	0	1
6th "	..	7	1	0
7th "	..	4	1	3

The second solution corresponds to the empirical answer given in the first instance.

The Young Man's Age

During one of the University vacations, a young man went to spend a few days with the parents of a friend. He was asked, "How old are you?" and replied, "The day before yesterday I was 19, and next year I will be 22."

During which vacation, and on what day could he have given this answer?

Solution.—During the Christmas vacation, and on January 1, his birthday being December 31.

On the eve of December 31 he was still 19; by January 1 he was 20, in the year when he was speaking he would be 21 on December 31, and, the next year, 22 on December 31.

What is the Engine Driver's Name?

In a train there are an engine driver, a stoker and a guard, who are named Smith, Jones and Robinson, but it is not known which is called which.

There are also three passengers, Mr. Smith, Mr. Jones and Mr. Robinson.

Mr. Robinson lives in Leeds.

The guard lives halfway between Leeds and Sheffield.

Mr. Jones earns $15,545 per year.

A passenger who is the nearest neighbor to the guard earns exactly three times more than the guard.

The guard's namesake lives in Sheffield, and Smith has beaten the driver at billiards.

What is the engine driver's name?

Solution.—The passenger who is the nearest neighbor to the guard (who lives halfway between Leeds and Sheffield) is neither Mr. Robinson, nor the guard's namesake.

On the other hand he is not named Jones. For Mr. Jones earns $15,545 a year, a quantity which is not divisible by 3. Mr. Jones cannot earn three times more than the guard.

The answer is that the passenger, since he is neither Mr. Robinson nor Mr. Jones, is Mr. Smith; the guard's namesake is Mr. Jones, and the guard is also named Jones.

It is left to attribute the names Smith and Robinson to the engine driver and the stoker. But Smith has beaten the stoker at billiards.

Thus, since Smith is not the stoker, Robinson must be, and the engine driver's name is Smith.

An Awkward Question

Write 23 with only 2's, or 45 with only 4's, or 1,000 with nothing but 9's.

Solution.—

$$22 + \frac{2}{2},\ 44 + \frac{4}{4},\ 999 + \frac{9}{9}$$

The Height of the Staircase

The height of a staircase is between 15 and 20 feet. Without reaching the top, you climb half the steps, then a third of the remaining steps and finally an eighth of the rest. Each step measures 8 inches in height. What is the total height of the staircase?

Solution.—You climb half of the steps, half are left. Then you climb a third of the rest, or $\frac{1}{2} \times \frac{1}{3} = \frac{1}{6}$. There remains: $\frac{1}{2} - \frac{1}{6} = \frac{1}{3}$. Then you climb $\frac{1}{8}$ of the remainder, but have still not reached the top.

This last information shows that a third of the number of steps is a number divisible by 8. Therefore the number of steps is divisible by $8 \times 3 = 24$. The number of steps in the staircase is 24 or a multiple of 24.

The height of the staircase, being between 15 and 20 feet, can only be: 8 inches $\times$ 24 = 16 feet.

The Successive Remainders of a Series of Divisions

What is the smallest number which when divided by 10 gives 9 as a remainder, divided by 9 gives 8 as a remainder, divided by 8 gives 7 as a remainder, etc., divided by 2 gives 1 as a remainder?

Solution.—If N be this number, the problem can be set out thus:

N = Multiple of 10 — 1
N = " " 9 — 1
N = " " 8 — 1
. . . .
N = " " 2 — 1

Therefore:

N + 1 = multiple of 10
N + 1 = " " 9
N + 1 = " " 8
N + 1 = " " 7
. . . .
N + 1 = " " 1

We have thus to find the lowest common multiple of the ten first numbers, which gives:

$$2^3 \times 3^2 \times 5 \times 7 = 2{,}520$$

The number required is therefore 2,519.

A Distribution of Pencils

A schoolmaster had N pupils registered for the beginning of term in September.

Each pupil was to receive a number of pencils equal to twice the number of his fellow-pupils reduced by 35.

2,150 pencils would remain over for the school stock.

At the beginning of term 10 pupils failed to arrive.

Each of these had received the same number of pencils and there remained 450 for the school store.

1. How many pupils were registered? How many pupils were present? How many pencils did each of these receive?

2. How many pencils would each pupil present have had to receive in order that none should be left over for stock?

Solution.—Let x be the number of pupils registered, y the number of pencils to be given to each of the x pupils, z the number of pencils given to each of the x — 10 pupils.

$$x\,[2(x - 1) - 35] = xy$$

$$(1)\ 2 \times x^2 - 372 = xy$$

Number of pencils given to the x — 10 pupils = (x — 10) z.

The difference between (x — 10) and xy is 2,150 — 450 = 1,700.

$$(x - 10)z - xy = 1{,}700$$
$$xz - 10z - xy = 1{,}700$$
$$(2)\ xz - 10z - xy = 1{,}700 + xy.$$

By subtracting (1) from (2) we obtain:

$$xz - 2x^2 + 37x - 10z = 1{,}700.$$

The first number should therefore be divisible by 100

$$x = 100.$$
$$100z - 20{,}000 + 3{,}700 - 10z = 1{,}700$$
$$90z - 16{,}300 = 1{,}700$$
$$90z = 1{,}800$$
$$z = 200.$$

On replacing x by 100 in the equation (1), we obtain:

$$20{,}000 - 3{,}700 = 100y$$
$$100y = 16{,}300$$
$$y = 163.$$

Hence:

1. Pupils registered: 100; pupils present: 90; number of pencils distributed to each of the 90 pupils: 200.

2. To distribute the total stock of pencils it would have been necessary to give 265 pencils to each pupil.

The Animals in the Zoo

A zoo contains antelopes, snakes and rhinoceroses. The number of horns equals half the difference between the number of hoofs and the number of heads. It also equals the product of the square of the number of snakes and the number of antelopes, plus the number of antelopes, less the product of the number of rhinoceroses and the number of antelopes less half the product of the number of snakes by the number of rhinoceroses. The number of horns multiplied and the number of snakes equals twice the product of the number of antelopes and the number of snakes, plus twice the product of the number of antelopes and the number of rhinoceroses. How many animals of each kind are there?

Solution.—Let x, y and z be the number of antelopes, snakes and rhinoceroses. The horns, hoofs and heads number respectively: $2x + z$, $4x + 4z$, and $x + y + z$. We have the three following equations:

$$2x + z = \tfrac{1}{2}\,[(4x + 4z) - (x + y + z)]$$
$$2x + z = xy^2 + x - xz - \tfrac{1}{2}\,yz$$
$$y\,(2x + z) = yx^2 + 2xz.$$

After simplifying, we have:

$$x + y = z$$
$$\frac{2x\,(y^2 - 1)}{2\,(x + 1) + y} = z$$
$$2x = y.$$

Which gives:

$$2x\,(2x^2 - 3x - 2) = 0.$$

The only acceptable positive root is $x = 2$.

In the zoo there are, therefore: 2 antelopes, 4 snakes and 6 rhinoceroses.

A Paradoxical Equality

Show that: $2 = 1.$

Solution.—Clearly we have:

$$2 - 2 = 1 - 1$$

or:

$$2\,(1 - 1) = 1 - 1.$$

On dividing the two sides of the equation by $(1 - 1)$ we obtain:

$$2 = 1.$$

This paradoxical result arises because the division has been made by a quantity equal to zero.

In a general manner it can be shown similarly that any number N is equal to another number N^2.

We have in fact:

$$N - N = N^1 - N^1$$
$$N\,(1 - 1) = N^1\,(1 - 1).$$

Dividing by $(1 - 1)$ we obtain:

$$N = N^1.$$

CHAPTER V

Animal Calculators

MODERN WORKS on animal psychology have shown that up to a point many animals can estimate sizes and quantities. Thus, Rothe mentions a horse that could distinguish easily between one lump of sugar and two lumps of sugar, but not between two lumps and three. Werner Fischel trained birds to find grains of corn in boxes which he had marked with dots. The more dots on the box, the more grains of corn it contained. A pigeon could distinguish 2 dots from 1 dot, and 3 from 1, but not 3 from 2. A goldfinch could distinguish 4 dots from 2 dots, 6 from 3, 12 from 6, 16 from 8, 18 from 9, but could not tell the difference either between 5 dots and 4 dots or between 10 dots and 6.

Katz and Revesz trained chickens to peck up only every other grain of corn in a row, and then to take only one grain in three, the others having been stuck to the ground in experiments designed to establish the desired reflex. But their attempts failed when they tried to make the hens peck up only one grain in four.

Professor Koehler trained birds to eat only a certain number of grains of corn in succession. A small machine poured out the grains one by one at regular intervals, and the birds stopped when they had swallowed the selected number. They were also trained to open boxes and to take only the number of grains they were allowed. The pigeons could "count" up to five, the jackdaws up to six and the crows and parrots up to seven.

A strange transposition of acquired ideas took place in the following experiment: A bird had been taught to open a box marked with three dots, but not to touch the box next to it, marked with four dots. Professor Koehler replaced the dots with live worms on the lid, and was surprised to see the bird hesitate a moment, and then decide not to open the box on which the three worms had been placed, but to eat these worms and ignore the four worms on the other box.

Experiments analogous to these can be carried out by an appeal to the sense of hearing. Experimenting with a macaw, Woodrow struck a number of blows with a hammer at intervals of a second and a half; the animal had to go and find food when it heard the longer of the two groups of sounds and remain motionless when it heard the shorter. In these conditions the macaw learned to distinguish one hammer stroke from three and three from four, hesitated between four and five, but regularly confused five and six strokes. Thus the animal could differentiate between the durations only if their ratio was sufficiently high.

Instances can be cited of birds which notice the disappearance of eggs from their nests or animals which notice the absence of their young.

If one egg in four is taken from a bird it will continue to sit, but if two out of four are taken it will abandon its nest. In the same way, if one baby field mouse is removed from a litter, the mother seems to pay no attention, but if two are taken she goes to look for them.

We could multiply these examples. In all cases the conclusion reached is that the animal reacts to a global evaluation of quantities but that it has no precise sense of number. Its psychic powers in fact are probably analogous to those which we ourselves experience when we compare two different quantities of the same kind without counting them, two groups of identical objects for example, or make rough estimates of unequal durations of time.

But, outside these classic and well-established facts, are there other conditions in which an animal could act as a genuine calculator? In other words, just as among men there are lightning calculators with faculties considerably beyond the mathematical capacities of ordinary mortals, are there gifted animals capable of carrying out complex arithmetical operations, such as additions, subtractions, multiplications, raising to powers and extractions of roots, and whose mathematical capacities are in consequence infinitely greater than the very rudimentary type of the species in general?

Some experimenters, among whom von Osten, Krall, Frau Moekel and Mme. Borderieux can be cited as the best known, have made such a claim. They consider that animals, especially horses and dogs, can be taught to count, if suitably educated. Let us examine the results they have obtained, so that we can consider how these should be interpreted.

Hans, the Berlin horse

"In 1900," writes Professor Claparède in the *Archives de Psychologie de Genève,* "Wilhelm von Osten, of Berlin, undertook the education of a Russian stallion named Hans. After he had familarized it with various accepted notions, such as right, left, up and down, lessons in calculation were begun by the intuitive method. Hans was brought before a table on which first one, then two and later several small ninepins had been placed. Von Osten knelt down beside Hans, pronounced the corresponding numbers and made the horse strike the ground with his hoof as many times as there were ninepins. Soon the ninepins were replaced by figures written on a blackboard. The results were surprising. The horse was soon able not only to count (by striking the number of blows required) but also to carry out genuine calculations and solve little problems unaided. . . . But Hans could not only count; he could read, he was a musician and could distinguish harmonies from discords. He also had an excellent memory; he could give the date for each day of the week. In short he performed operations up to the capacities of a normal fourteen-year-old schoolboy."

A good deal of lively scientific argument ensued and in 1904 a first committee of enquiry was set up. It was composed of Herr Strumpf and Herr Nagel, professors of psychology and physiology at Berlin University, the director of of the Zoological Gardens, a circus manager and a sprinkling of veterinary surgeons and cavalry officers. Its report concluded that no tricks had been used by von Osten, since the experiments succeeded even in his absence, but it gave no indication of a possible explanation for the facts observed.

A second committee, composed of Herr Strumpf and his associates, came to the conclusion that the horse could count, read or calculate only when the persons present knew the answers to be obtained. At this stage Herr Oskar Pfungst, a graduate of the Berlin School of Psychology, carried out a detailed research which led him to assert that the horse possessed no intelligence at all, that he could not recognize either letters or figures, that he did not know how to count or calculate, but simply obeyed the imperceptible unconscious movements of the heads or eyes of people near him.

Von Osten, discouraged but not convinced, abandoned

his experiments and died in obscurity in 1909 at the age of 71.

The Elberfeld Horses

Von Osten and his works seemed to have passed out of memory when Karl Krall, a wealthy merchant of Elberfeld revived the investigation. He had acquired the horse Hans in 1908 and now added two Arab stallions, Muhamed and Zarif, and began to teach them. Lessons lasted for an hour and a half to two hours daily for each animal. First they were taught the figures up to 9, then tens were explained to them and they were taught to tap these out with the left hoof, the right hoof being reserved for units. As had been the case with Hans the progress made was astonishingly rapid. Less than two weeks after his first lesson, Muhamed, who showed himself to be the more "mathematical" of the two new pupils, was doing simple additions and elementary subtractions correctly. He mastered the meaning of the signs + and —. Four days later he started multiplications and divisions. At the end of a few months he knew how to extract square and cube roots, soon afterwards he learned to spell and read by means of a conventional alphabet invented by Krall, in which each letter or diphthong was represented by a number. Here is the alphabet, as reproduced in Krall's work, *Denkende Tiere*.

The figures aligned horizontally indicate the taps to be

	1	2	3	4	5	6
10	E	N	R	S	M	C
20	A	H	L	T	Ä	CH
30	I	D	G	W	J	SCH
40	O	B	F	K	Ö	Y
50	U	V	Z	P	Ü	
60	EI	AU	EU	X	Q	

made with the left hoof, and the significant figures (not the numbers) arranged vertically, those to be made with the right hoof. Thus to indicate the letter P the horse would have to make four taps with the left hoof and 5 taps with the right hoof. By this process Zarif was able at the end of four months to spell out words spoken before him which he had never seen written. A somewhat curious fact is that he often left out the vowels of words, but when he was asked to add the missing letter or letters, he gave them correctly. Thus for the word Zucker (sugar) the following spellings were obtained: *zkr, zukr, zuker,* etc.

In time Herr Krall's stable was enriched by four more horses: a little pony, Hänschen, two Arab stallions, Amasis and Haroun, and a blind horse, Berto.

Naturally these events attracted a crowd of visitors and scientists to Elberfeld and several of these applied themselves to checking the authenticity of the phenomena and investigating their probable mechanism. Among these scientists was Dr. Edinger, an eminent German neurologist; Drs. H. Kramer and H. E. Ziegler of Stuttgart; Dr. Paul Sarasin of Basle; Professor Ostwald of Berlin; Professor A. Besredka of the Pasteur Institute in Paris; the well known Swiss psychologist Dr. Ed. Claparède, professor at the University of Geneva; Professor Schöller; Dr. Gehrke of Berlin; Professor Goldstein of Darmstadt; Professor William Mackenzie of Genoa; Professor R. Assagioli of Florence; Dr. Hartkopf of Cologne; Dr. Freudenberg of Brussels; Dr. Ferrari, professor of neurology at the University of Bologna, etc.

All were unanimous in ruling out the possibility of trickery of any sort. Moreover the scientists often worked alone with the horses and put questions to them which had been prepared in advance.

"Professor Assagioli and myself," wrote Dr. William Mackenzie, "were able to work with the pony Hänschen on several occasions without anyone else present, either Herr Krall or his grooms. While doing so we obtained a number of immediate and accurate replies to problems which we wrote on a slate, such as 33 + 44, 42 + 23, etc.[1]

"The questions were generally set out on a board, after which everyone went out and followed the movements of the animal by looking through small glazed holes cut in the door.

[1] The pony Hänschen was at that time a recent acquisition and not yet as advanced as Muhamed and Zarif.

"Thus," Professor Mackenzie reports, "the solution of the root $\sqrt{1874161}$ was given by Muhamed by means of hoof taps while were in the courtyard and observing the horse through small holes in the stable door."

Problems were also set by third parties and placed in sealed envelopes. This was done by Dr. Hartkopf of Cologne. He obtained immediate answers to the following problems:

$$\sqrt[3]{13{,}824} \qquad \sqrt[3]{29{,}791} \qquad \sqrt[3]{103{,}823}$$

As the solution to the third question the horse gave, first, 57, which was wrong, then immediately afterwards, 47, which is correct. Similar results were obtained by the same method by Dr. Haenel, by H. Overbeck and some other investigators. Dr. Haenel for example took from an envelope a problem of which he did not know the solution:

$$\sqrt[4]{7{,}890{,}481}.$$

Muhamed replied: 53. The doctor looked at the back of the paper: again the answer was correct.

"One fact repeated itself fairly often," writes Professor Claparède. "This is that the horse often makes a mistake in answering a question put to him; then another question is put and instead of replying to the second question it is the answer to the first one that turns up."

Here is an example of this: in the course of an experiment Professor Claparède put the following problem to Muhamed:

$$\sqrt[4]{9{,}834{,}476}$$

The animal gave several wrong answers. Claparède then put another question of the same kind and Muhamed gave the answer 56, which is the solution to the previous problem.

Quite the most remarkable thing about the Elberfeld horses was that they gave their answers immediately, or after an interval of only a few seconds. Professor Claparède notes this odd fact in the following passage from his report:

"The cube root $\sqrt[3]{5{,}832}$ was written on the board and given correctly (18) in a few seconds. The operations:

$$\sqrt{15{,}376} = 124 \text{ and } \sqrt[3]{456{,}776} = 26$$

were given correctly in about ten seconds. Herr Krall and his groom had left the room. None of the persons present knew the answers."

Thanks to the special alphabet the horses "conversed" with visitors and sometimes answered questions put to them with a certain humor.

Muhamed was shown the portrait of a young girl whom he did not know. "What is this?" he was asked. "A girl." "Why is it a girl?" "Because she has long hair." "And what does she not have?" "Mustache."

He was then shown a picture of a man without a moustache. "What is this?" "A man." "Why is it a man?" "Because he has short hair."

One day Zarif, Muhamed's companion, spontaneously made the following remark to his master: "Albert has beaten Hänschen."[1]

Another time he said: "Hänschen has bitten Kama."

The unfortunate Elberfeld horses were requisitioned in 1914 and ended their lives on the battlefield.

The Mannheim dogs, Rolf and Lola

Krall's experiments soon gave rise to a veritable "epidemic" of animal calculators in Germany. Efforts were made to educate dogs and cats, and even an elephant. The results were generally poor, but at Mannheim Frau Moekel achieved complete success with her dog Rolf. Here is her own account of the animal's first manifestations of intelligence: "One day at about noon I was sitting near the children, engaged in the tedious task of helping them with their homework. Our little Frieda, who is amiable and lively but a trifle scatterbrained, was obstinately failing to grasp the solution to the problem of 122 + 2 when in a momentary impatience I administered a gentle slap. At this moment the dog, who was lying under the table, gazed at us with such big eyes that I said: 'Frieda, look at Rolf. He looks as if he knew the answer.' Rolf came up to me, sat down at my side, and looked at me with his big eyes. I said: 'Rolf, what do you want? Do you know what 2 and 2 are?' Whereupon, to my great amazement, he tapped my arm four times with his paw. Our eldest child at once suggested asking the dog how many 5 and 5 make. The answer followed immediately, ten taps with his paw. Continuing these tests that evening we found that the animal could solve simple problems of addition, subtraction and multiplication correctly."

If Frau Moekel is to be believed Rolf had learned all these operations of his own accord. Soon, too, he invented a conventional alphabet for himself.

[1] Albert was the groom.

"During a calculating session I asked Rolf earnestly if he wished to learn to talk," Frau Moekel writes. "The animal replied 'yes' with great eagerness.[1] I then asked him: 'Rolf, would you like to make an alphabet for yourself?' Again he answered 'yes,' very emphatically. I continued: 'Rolf, pay great attention now. I am going to tell you the letters of the alphabet and you will tell me the number which should stand for each letter. What do you give me for A?' At once he replied: '4.' 'Next for B?' Reply: '7,' and so on. I noted the numbers given Rolf and the next day I found to my amazement that the animal had remembered them."

This alphabet was composed thus: 1. of 24 tap-signs corresponding to 24 letters; 2. Of 5 conventional signs, corresponding to five words: *ja* (yes), 2 taps; nein (no), 3 taps; *müde* (tired), 4 taps; *Gasse* (street, go out in the street), 5 taps; *Bett* (bed, go to bed), 7 taps.

From this time on Rolf was able to hold actual conversations with his mistress or with visitors. One day a bunch of flowers was brought in and Rolf was asked to give the number of flowers it contained, after everyone had counted them. Some said 18 flowers, others 16. Rolf indicated that there were 14. The bouquet was then taken apart and the following questions were put to the dog: "How many mauve flowers?" Answer: "1." How many yellow flowers?" Answer: "4." "How many blue flowers?" Answer: "9." The recapitulation proved Rolf to be right and everybody else wrong.

To Dr. Mackenzie of Genoa, Rolf made this unforeseen reply. The doctor asked him: "What is autumn?" "Apple time," replied the animal. One day Frau von Schweitzerbarth, a lady of mature age and ancient nobility, asked him what she could do to please him. He at once replied: "Wag your tail."

Professor Ziegler has claimed that Rolf's replies revealed the mentality of the animal and not that of his mistress.[2] This is in fact what would be suggested by the answers we have just quoted, as well as by the following. The professor showed Rolf a postcard representing a crocodile. When brought to Frau Moekel the dog at first answered: "What does it matter what it is on the stupid card, give me something to eat instead." When they per-

[1] Probably by tapping his paw an agreed number of times.

[2] Psycho-analysts have revealed that our unconscious psychism is indeed immense and disturbing.

sisted he said: "Odd beast, don't know exactly." At another sitting Professor Ziegler showed the dog a drawing of some black hens, without telling Frau Moekel. When questioned the dog replied: "I can't eat those black creatures in the drawing, but you are gorging yourselves all day." By this rude answer the dog is supposed to have shown his annoyance because the guests were eating cakes with their coffee without offering him any.

Rolf's mathematical capacities equalled and even surpassed those of the Elberfeld horses. For instance he could solve complicated problems involving cube roots of the following type:

$$\sqrt[3]{1,331} + \sqrt[3]{1,000}.$$

Independent tests were made in 1913 by Dr. Oshausen, of Hamburg, who declared himself convinced of Rolf's calculating ability. Frau Moekel died in 1915 and from that time on Rolf's faculties declined. This was noted by Dr. Ziegler in 1916 and by Robert Lutz in 1919.

"We no longer get the astonishing answers which seemed more human than animal," Mme. Carita Borderieux writes. "Instead we get rather those which any intelligent dog could produce if he knew how to express them in our language." In any case the dog was approaching the end of his career, since he died in 1919.[1]

Rolf left four descendants: Ilsa, Lola, Heinz and Harras. The best known of these was the bitch Lola.

Born in 1914, Lola had had the beginnings of a mathematical education from Frau Moekel, but without any great success until 1916, when she was entrusted to a Frau Kindermann. She was soon able to get the dog to tap her right hand as many times as lumps of sugar were offered. Then progress was made with the addition of figures; $2 + 3, 1 + 3, 2 + 6, 4 + 5$, etc., then with a child's counting frame. After ten lessons of half an hour each Lola carried out small additions quite easily and gave the answers by tapping out the tens with the left paw and the units with the right. Nine days were sufficient for learning multiplication. Finally subtraction was tackled, then division and the reading of letters and words, with such success that at the end of a month, representing about fifteen hours of study, Lola could count and read.

Frau Kindermann wished to find out whether the animal could learn the notes of music; she played them on the

[1] The same phenomenon was noted with the American horse Black Bear which lost its abilities at the death of its master.

piano, giving the names successively, do, re, mi. . . . In a few lessons the dog knew the notes. Lola was then taught chords and learned to recognize the number of notes in each of these.

Lola died in 1920 but left a puppy, Awa, who was soon given instruction by Professor Ziegler. Awa learned to do additions, subtractions and multiplications and to use a conventional alphabet, with which he could answer simple questions. But he made little progress in division and was never able to tackle the extractions of roots which his grandfather, the illustrious Rolf, had mastered. The animal tapped his master's palm, and this is clearly a weak point in the method, which was however employed by other psychists later. The professor tried using a board, but without success. This negative result, he said, was due to the fact that a board is an inanimate object, while the taps on the hand keep the dog in personal contact with its master. As we shall see later, the real reason is quite different.

The dog Zou of Paris and his imitators

Soon other calculating dogs made their appearance in Germany and elsewhere: Seppl, Senta, Buzi, Ali, Lady, Sam, Bim, Tommy, etc. A description of their performances would lead us into needless repetition. France, however, the land of Descartes and of robust common sense, a sense perhaps at times a little too down to earth, did not follow this movement which led numerous experts in a number of countries to make these unusual experiments in animal psychology.

It was now however that Mme. Carita Borderieux, a woman of great talent, a distinguished psychist and editor of the periodical *Psychica,* who was personally acquainted with the principal protagonists in the new studies, Krall, Frau Moekel and Frau Kindermann, and who had read the already abundant literature devoted to "thinking animals," began in her turn to educate a young dog, Zou, whose mother was an English fox terrier and whose father a French sheep dog.

"It was on March 28, 1925," Mme. Borderieux writes, "when Zou was just fourteen months old that I decided to give him his first lesson. After lunch, while there remained on the table a plate of cakes which the dog specially liked, and Zou was alone with me, I installed him on a chair, put out my right hand and rested it on the table, saying to him: 'You see Mamma's paw, it is making one.' Then putting out my hand to him I said: 'Make one.'

"Obediently, Zou lifted his paw and put it in my hand.

"I carried out the same demonstration for 2 and Zou imitated me, then for 3, 4, 5 and 6. We stopped at this figure for a time. Then I began again: 'Listen, 1, 2, 3, 4, tap 4.' The dog tapped the right number. Next I varied the number of taps: 'Tap 4, 2, 1, 5' repeating these aloud; the result was right each time. . . . In two days (ten minute lessons) Zou learned to count by looking at my fingers.

"On April 10 I showed him figures for the first time. I had obtained a tear-off calendar and on that Monday I made the dog read the first six numbers. I placed the card bearing the figure 1 against the box of cakes (for these had always to be kept handy). I said to Zou: 'Look!' and I made the gesture for 1 with my hand; then I showed him my thumb; 'Make one!' The animal made one tap with his paw. I repeated the operation for the first six numbers, followed obediently by my pupil, who looked at the figures for a long time. I then took a plate of figs, spread them out, and made the gesture of counting with my fingers, but without saying a figure aloud. Then I said to the little dog: 'How many?' He made 14 taps with the same paw; it was correct . . .

"As the tens were now to be indicated by taps with the right paw and the units with the left, in order to avoid too great a number of taps, my little pupil had considerable difficulty in getting this change-over into his head. On April 22 I taught Zou subtraction. How I explained it to him, I really don't know, and I think that neither you nor I, if we had known nothing about it, would have understood the involved explanation I gave him.

"Fortunately I gave examples. I showed him 7 with my fingers, lowered 3 and said: '7 minus (emphasizing the word) 3 leaves 4.' Since then my pupil has made no mistake over the words plus and minus. . . . Moreover Zou always shows a great preference for subtraction." Excelling in the domain of figures, Zou remained only passable when he was taught the alphabet, so that he could answer questions only by "yes" or "no."

Zou began to be shown off in public in 1923 and until his death in 1935 he took part in numerous exhibitions both in France and elsewhere. I saw him on several occasions, first in public, then in private at Mme. Borderieux's home. He replied correctly to questions I put to him, such as: "How many people are in this room?" "How many are 8 plus 3?" "9 minus 4?" etc.

Mme. Borderieux's example was soon followed: many other French dogs, Diane, Zou II, Androcles, etc., were educated by her method, a method of persuasion and gentleness and based on the concept that the animal "soul" is very little different from the human "soul." It is very well expressed in the following advice given by Frau Mathilde von Freytag-Lovinghoven, of Weimar, who taught the dogs Lumpi and Kurwenal, by following Mme. Carita Borderieux's instructions.

"Do you wish to condition a dog to the language of numbers," she writes, "so that he may learn to express words? Then above all you must show the animal warm affection, and with this a sympathetic and friendly understanding of his character, which is strongly marked in all animals and particularly so in dogs of all breeds. Any irritability, anxiety, impatience, unkindness or inattention on the teacher's part will have repercussions in the animal and reduce his zest for the work. This absolute confidence of the animal in his master, or the teacher his master has given him, must never be shaken. In return the teacher, on his side, must show complete confidence in the dog. If a doubt is expressed: can the dog understand this or that? it is like an insult to the animal, and not only does it not encourage him to work, but it annoys and mortifies him. The dog understands nearly every word, but if there is one he does not understand, his gift of combination and his penetration of ideas are such that he can manage without exact comprehension of this word.

"This then is the beginning; later, with a dog that has been taught, the intelligence functions more or less with the same freedom as with us. . . . The educated dog always works willingly; he is pleased and proud of his knowledge and does not get tired, at least in appearance. However one should guard against his willingness. Here the affectionate attention of the teacher must also intervene and protect the pupil from an overstrain which often shows itself very suddenly among dogs and whose effects can be dangerous and lasting. Most dog pupils learn with a genuine enthusiasm. This can be fostered by caresses and affection; their errors and faults must be explained to them gently."

CHAPTER VI

How Animals Do It

TO EXPLAIN the feats performed by animal calculators, five theories can be considered: the mediumistic theory, the telepathy theory, the theory that the animals really do solve the problems they are set with their own intelligence, the theory of signals unconsciously made by the operator or by others present, and finally, training.

Mediumism

The mediumistic theory has been maintained by César de Vesme, an eminent psychist. He has compared the animal calculator to the spirit medium which supplies paranormal information by the table instrument; the animal's paw taps the ground as does the table leg. M. de Vesme believes that a mental automatism resembling that of a medium exists among calculating animals.

"It is well known," he writes, "that the different personalities which reveal themselves by typtology have a special way of tapping, so much so that when they follow each other in succession they can often be identified by habitués even before they are named."

Similarly each of Herr Krall's horses had its own way of making hoof taps. Dr. Mackenzie observed and noted this.

"The little pony Hänschen taps in a very special way. This curious little animal, which is not much bigger than a large dog, makes its taps almost by scraping the ground and nods its head to the same side as it does so. When it has finished it turns round to see the effect on the audience; often when there is no applause, because the answer has been wrong, it starts its scrapings again and usually gives the correct answer."

Similarly it has been noticed that the table taps, unequal and hesitant when the spirit entity is having difficulty in giving an answer, become clear and definite when the entity believes itself sure of what it is saying. Here is what Dr. Ferrari writes on this point:

"It should be added that several observers have noticed that the horse taps differently when it is uncertain and when it is sure of giving the correct solution. In the second case he is more energetic."

The analogies between the animal calculator and the medium can be extended. It is known that one of the characteristics of the automatic writing of mediums is that they are often produced in reverse. This is what is called "mirror writing." The same thing happens with the tables; the letters of a word are often mixed up or arranged in inverse order. Now Hänschen sometimes reversed the numbers he was given to read: 45 instead of 54; 65 instead of 56. Dr. Ferrari notes:

"The errors are frequent but they are usually logical, in the sense that they are made up of the figures which ought to compose the number representing the required solution, but they are arranged in an erroneous fashion, without method, up to the moment when the horse finds the right answer, either spontaneously or after a crack of the whip.

"Moreover, just as is the case with mediums, the animal calculator, after it has solved complicated problems, is sometimes incapable of replying to quite simple questions such as giving the number of onlookers. Finally, as with the lightning calculators, the animals' answers are almost instantaneous."

"Thus," concludes M. de Vesme, "we are led to the theory that there are subconscious paranormal faculties among beasts which not only render these creatures, which we tend to regard as our 'lesser brothers,' more like ourselves, but even make them our equals in potentialities. An equal intelligence seems to be distributed among all creatures if we take into account the supranormal consciousness or *subliminal self,* an intelligence both superior and mysterious, which Plato called 'divine,' and which those strange geniuses the Moslem Averroes and the Jew Spinoza placed at the base of their pantheistic systems."

To this theory of mediumism the recent theory of "polypsychism" advanced by Professor William Mackenzie and by René Warcollier, president of the International Psychic

Institute, can be related. According to this notion calculating animals possess a sort of "psychic agglomeration," an entity which participates in the thought, will and mental power of those who generate it. "Poly-psychism," writes Dr. Mackenzie, "necessitates an interaction between the component parts of each organic system and a fusion of the psychic elements of these parts to the whole. Each whole is not a sum but a product. It is absolutely new, and possessed of vital force."

It results from this concept that the answers given by "thinking" animals do not actually come from them, but derive from a psychic union, so that the eminent professors who journeyed to Mannheim and Elberfeld to study the animal "calculators" themselves furnished the answers which they had come to seek, but unknowingly and without any question of transmission of thought. It was the increasingly intellectual quality of the examiners which enabled the animals to progress with such ease from the elements of arithmetic to the solving of complicated problems like the extraction of cube and fourth roots.

These theories, whether they involve mediumism or poly-psychism, are obviously very attractive, but apart from the fact that they are difficult to demonstrate, they do not supply a useful working hypothesis, on account of their purely psychic character, so that it is well to consider whether the problem of animal calculators cannot be approached from other points of view which could guide and facilitate future experiments. The theory of telepathy which we shall now put forward and discuss, seems, *a priori,* to fill this need.

Telepathy

The theory of telepathy is frequently supported by the owners of calculating animals and by some eminent psychists. The suggestion is that the experimenter, who usually knows the right answer, unconsciously transmits it to the animal by telepathy. When he does not know the solution a subconscious mechanism comes into play in his mind, in the way in which it probably functions among lightning calculators, and the solution is then communicated to the animal by extra-sensory means as in the previous case. "Although transmission of thought between man and man is a rare and special faculty," writes Karl Krall, "that between man and an animal, at any rate in certain conditions, succeeds easily, as I have found myself."

On the same lines the Russian psychologist Bechtereff asserts in his treatise *Collective Reflexology,* that an instant psychic induction from man to animal is possible.

"In all I have carried out between 50 and 75 experiments with dogs," he writes. "They were done with three dogs, Lord, Piki and Dezi. The will power of those dogs was first reduced by training, so that the animal selected for the experiment would remain calm and able to concentrate on the task to be performed. Thus under the influence of mental suggestion by the experimenter, Lord (a St. Bernard) barked as many times as he was mentally ordered to do. The number of barks so imposed must not exceed 9. Otherwise the experiment does not succeed, because the dog then continues to bark automatically. I have noticed the same phenomenon with the bulldog Dezi. These experiments were successful even when the induction was practiced from an adjoining room with the doors closed.

"The fox terrior Piki performed more complicated tasks. The experimenter would suggest for instance that he should jump on a table near the wall, from there to a smaller table, then stand on his hind legs and touch a portrait on the wall with one of his paws. Another experiment: to jump on a chair in front of a piano and tap the right side of the keyboard with his paw. Third experiment: to run into an adjoining room (where some paper balls had been set out) and go to one ball which had been indicated mentally.

"For the experiments with Piki and Dezi precautions were taken to avoid any suspicion that the animals carried out their tasks under the influence of any signs consciously or unconsciously produced by the experimenter. In the latter tests the experimenter was the only person aware of the task to be performed. He was separated from the dog, he did not touch it, and he operated only at a distance, solely by concentrating his thought upon the animal's task. Then he withdrew. The mental suggestion was carried out nevertheless with satisfactory precision. The suggestions made to Dezi were obeyed even when the inductor was himself in another room with the doors closed."

On his side, Professor Edinger, author of some sound books on the nervous system and psychic manifestations of vertebrates, declared:

"It is certain that the Elberfeld horses could actually read, count and write and that their master avoided doing or saying anything which could have served as an indication. Here we have something significant; either the soul of the animals is hereby revealed or else, more probably, there is a mysterious transmission of thought."

Mme. Carita Borderieux also acknowledged the possibility of communication of thought between man and animal.

"I compare the calculating or talking animal to a somnambulist," she wrote. "Like the somnambulist he finds himself in a state of *rapport* with certain persons, with his teacher especially. Just as the best of somnambulists are useless with certain interrogators, so an exceptional subject, put to sleep by one hypnotist, will display no clairvoyance when put to sleep by another."

That a kind of telepathy consisting of a transmission of rather vague impressions by "waves" of an electromagnetic nature can occur between animals of the same species such as ants, bees, termites, or between higher animals like cats, dogs, horses and sometimes even between these last and man, is perhaps not completely ruled out, but that a precise command can be accurately transmitted mentally from a man to an animal seems to be quite impossible, despite the fact that Bechtereff's experiments seem to show otherwise; these will be criticized later on.

Moreover, if transmission of thought between man and animal did exist, what would be transmitted? Would it be figures, or words? But then this would presuppose the animal to have the very intelligence which it is precisely the object of the telepathy theory to explain away. In other words the theory of telepathy implies this other hypothesis, that the animal is capable of *thinking and judging in human terms,* and this is very difficult to admit.

No doubt it could be presumed that what is transmitted, where arithmetical operations in particular are involved, is simply an order to cease tapping with the hoof or paw, but then it would be necessary for the indication to tap with the left or right foot to be communicated at the same time, which is very extraordinary indeed. This idea could at least be given serious consideration if everything were reduced to giving an answer to a simple problem like $12 + 9 = 21$, but it seems impossible to apply this theory of mental suggestion to complex problems where the animal has to supply a whole series of figures, and where he has to carry out complicated actions or conduct virtual conversations with men.

Intellectual aptitudes

A third hypothesis is to suppose that the animal calculators do really carry out the arithmetical tasks set them and that the "talking" animals really converse. This was the opinion of Professor Ziegler, for whom the feats of these animals were a revelation.

"Now that I have assured myself," he writes in one of his works on psychology, "that the Elberfeld horses and the Mannheim dog are genuinely capable of reading and calculating, *by their own reasoning powers,* I consider it to be my duty as a scientist to support the claims made for these experiments by word and pen, although I do not disguise from myself the difficulty there is in making known to science discoveries which are in contradiction to accepted opinion."

By this concept we are thus let straight back to the notion implied in the telepathic theory, which is that the animals have a human intelligence. But this is a proposition which, apart from the facts we are dealing with here, is in no way compatible with normal animal behavior.

There can be no question here of considering the problem of animal instinct and intelligence in detail. Let us say only that instinct reveals itself as an extremely complex assembly of ready-made mechanisms, necessitating no individual apprenticeship, which in a given species reproduces itself always in the same form or in a form very closely allied to it. It appears further as an activity which seems to be adapted to certain useful ends, either in the life of the individual or in the life of the species.

"An instinct," writes William James, "is a faculty for accomplishing certain acts for certain specific objectives, without previous instruction in these acts."

From these definitions it is clear that the feats of "thinking" animals cannot be explained as instinctive acts.

Animal intelligence as distinct from instinct displays itself among the higher species only in rudimentary forms and in a rather sporadic fashion. In modern scientific research it is revealed by means of a series of tests: mazes for mice, boxes and sticks for monkeys, "planchette" tests, in which cats excel, etc.

The overall picture supplied by these tests enables what is usually described as the "mental level" of the animal to be fixed and the various species to be classified according to their greater or smaller aptitude for solving problems set to them. As a general rule it can be said that apart from instinct the three stages of animal intelligence are: 1. the purely reflex action which is concerned only with the present moment; 2. the conditioned reflex action, which assumes a memory and a minimum of complexity in the psychonervous associations; 3. reasoning, which is the projection of the present into the future.

Now it seems clear that genuine reasoning, the association of complex ideas which brings with it a picture of the future, is rather a rare act in the animal. On the other hand, what is often observed in animal behavior is an astonishing faculty of memory, reinforced by a keen sense of observation. Almost always it is after having seen an act carried out by one of its own kind, or by a man, that the animal is led to accomplish it. There are cases however in which the animal shows proof of genuine intelligence, whether in its normal life or in the course of laboratory experiments.[1] Thus George Leroy, who has made a long study of the habits of wolves, has frequently observed that during an attack on a flock the female wolf will show herself to the dog which she leads away by making him pursue her. The male then enters the enclosure and can carry off the sheep it selects without risk. In their collective hunting the "painted wolves" of Africa (Lycaon pictus), carnivores which combine some of the characteristics of the dog, the wolf and the hyena, are as cunning as their cousins of the colder regions. The leader runs some hundreds of yards ahead of the pack and forces the prey to run while also forcing it to turn so that it will be brought round towards the pack in the rear. The technique is very effective and the lycaons are particularly feared by grazing animals, especially antelopes.

The coatis, which are small carnivores, hunt in packs in Central and South America. They live on fruit, insects, lizards, young birds and also iguanas, of which they are very fond. When a troop of coatis is hunting an iguana, is divides into two groups. One group remains on the ground while the other chases the iguana in the trees. After

[1] It should be noted that the intellectual qualities observed among animals at liberty are often strongly tinged by instinct. In some of the examples quoted here the part played by instinct is quite clear.

a time the animal, frightened and exhausted by the pursuit, drops to the ground. At this moment the coatis which have remained below and have followed the reptile's movements, leap forward and seize it.

The fox, the badger and the otter easily avoid the traps which are ordinarily set at one of the openings of their lairs. The stag, hunted by a pack of dogs, tries sometimes to ring the changes with another stag whose presence in the area he knows. The hare employs the same stratagem; when he has been chased for some time he will approach another hare in repose, push him out of his burrow and take his place; the dogs find themselves chasing a fresh animal which can often outdistance them. Sometimes, too, the hunted animal takes refuge in the midst of a flock of sheep, where he remains until the pack has lost the scent and abandoned the chase.

Rats too are noted for their cunning. La Fontaine showed in one of his fables how these rodents carry eggs. The method is so odd that one might think he had invented it; but no, the naturalist is here in agreement with the story teller.

"The fact," writes Robert Goffin, "has been confirmed by a reliable witness who assured me that he had seen two rats, one of which held an egg between his four paws while the other dragged his comrade by the tail; he watched them closely and was astonished to see the two thieves cross a yard as far as the end of a drainpipe, where the second rat stopped an instant to get its breath and then dragged his friend as far as the gutter at roof level."

Many authors have related a thousand and one examples of mischief by rats. Grosbois cites the case of a sack of potatoes ripped up and all the potatoes transported to a secret reserve; he also explains the manner now become classic, whereby rats empty bottles of oil and other liquids which please them. One of them gnaws out the cork and perches on top of the bottle, dips his tail into the neck and passes it to his partners who suck off the agreeable liquid, then take the distributor's place so that he in turn can benefit.

Marks of intelligence analogous to these, or of a slightly different kind, are also found among other animals; bears, elephants, lions, etc.

But it is above all in the laboratory that it is possible to perceive evidences of animal intelligence. A chimpanzee is shut in a cage; he is hungry; some bananas are out of his reach, but he can reach some sticks. If he uses a stick to pull the bananas through the bars of his cage, which he sometimes does, he is giving proof of intelligence. Monkeys will even use a short stick to draw towards them a longer stick which will enable them to reach the bananas. This act is clearly an intellectual one.

When a stick taken by a chimpanzee to draw the fruit towards him is obviously too short, it is generally abandoned without being used. But sometimes when he has a stick which is too short, the animal will take another which is as short as the first, or even shorter, place them together by their flat sides, and, holding them together with his hand, try again to reach the fruit. The method is clearly defective, but it is at least an attempt to improve the instrument. On the other hand if the two sticks can be inserted one within the other at their ends, the animal is capable of fitting them together in order to obtain an instrument which is long enough. We can recall here the following experiment carried out by Köhler with the monkey Sultan.[1]

"Two cane stalks, of approximately the same length, were placed at Sultan's disposal; the two stalks could be fitted one inside the other like lengths of a fishing rod. The fruit placed outside the animal's cage was at a distance such that it could not be reached with either stalk, but only by the length which would result from joining them together. Sultan began with a useless action: he pushed one of the canes outside the cage, as far as possible, and used the other to push it slowly and carefully in the direction of the fruit. He succeeded thus in establishing contact between the fruit and himself, but without being able to draw it towards him. It happened that in pushing he tried the cane he held in the opening in that which lay on the ground, but this fact did not make him join the two together.

"The observer then introduced his own forefinger, before the animal's eyes, into one of the canes, but without showing him the other. This maneuver had no effect. Sultan stopped making any further attempts and sat down on a box near the bars; then he got up, played with the canes in a distracted fashion, took one in each hand and suddenly, the two canes happening to be in position for joining, he pushed the thinner one inside the opening of the larger.

[1] Achille Urbain, director of the animal section of the Jardin des Plantes: *Psychologie des animaux savages.*

DHG

He at once ran to the bars and drew the fruit towards him with the double stick thus produced. When one of the canes fell, the monkey put them together again. This maneuver appeared to give him keen enjoyment; he drew all the fruits near the bars without taking the time to eat them and if the observer separated the two canes he at once put them together and continued to use them to bring towards him a number of different objects.

"In another experiment, with the objective at a greater distance, Sultan was given three canes of such dimensions that the two thicker ones could be put together with the two ends of the third. The monkey did not attempt to assemble the two thick canes; first he tried without success to use a double cane composed of the thin length and one of the thick lengths, holding the end of the former in his hand. Suddenly he drew the double stick towards him, turned it round, now having the thin end before his eyes, the other thicker end being in the air behind him; he then took the third cane in his left hand and joined it to the two others. With the help of this triple stick he reached the fruit and as this long instrument was difficult to handle he disconnected it by degrees as the fruit approached nearer to the bars.

"Sultan has never tried to assemble elements which could not be so adapted because of their dimensions or any other reason; thus when given two canes of unequal thickness he took the larger in his left hand, and thinner in his right hand and prepared to assemble them when suddenly he stopped, turned the thick cane round to examine its other end and then dropped them both. The thick cane ended at both ends in knots, without openings, and the monkey, having perceived this, had abandoned his attempt. As soon as the observer cut off one of the knots, the monkey at once made the join. In other experiments Sultan succeeded in introducing into a cane, to lengthen it, either a piece of wood or splinters of cane, whose diameter he reduced by gnawing them in order to facilitate the fitting together."

This is the kind of intellectual feat which can be observed among animals, particularly among the most intelligent of them, the monkeys. They are still no more than gleams of lucidity, full of interest no doubt and even moving, but still separated by an impassible gulf from the extraordinary mental faculties attributed to calculating and talking animals. Moreover, if these latter did in fact possess the conscious faculties which some authors attribute to them, the fact would be contrary to everything that human experience has taught us up to now about heredity and race. We know that mutations present a certain homogeneity and, although they work by "jumps", that they are in a sense regularly progressive (or regressive), yet we also know that the highest mentality evolves in an ordered manner and that it is only the development of social faculties to a higher degree.

No doubt it could be retorted that the higher signs of intelligence can manifest themselves among animals if they have the opportunity to be properly educated by human beings. But it is precisely those who have specialized in training animals who are unanimous in asserting that there is a limit, narrowly circumscribed, to the educability of each species and each individual, a limit which it is impossible to pass and which is thenceforward known and defined. Moreover, the "learned" animals learn too quickly to have had time to understand; they solve complicated problems as easily as simple ones and often more easily, they sometimes begin to tap the answers before they have even seen the problem set them. The nature of the mistakes they make is not that of the faults a genuine calculator could commit: it is primarily a transportation of figures, of mistakes consisting of a single unit; all mistakes which could easily arise from a confusion in the foot taps, or from the fact that this stops an instant too soon or too late. There is no further doubt possible; contrary to the widely-held theory, the animal calculators do not carry out the operations put to them; the talking animals do not know how to express themselves in German or in French, or in any other language.

Unconscious movements and training

And so we come to the only true explanation of the problem of the thinking animals: the theory of unconscious signals and training.

Where the Elberfeld horses are concerned Professor Claparède has noted that these were far from having the quasi-infallibility of human lightning calculators. The number of their errors was large and, a fact which appears to be paradoxical and which we have already explicitly noted more or less, the Genevan psychologist states that in a series of experiments he observed 11 per cent of correct answers for easy questions and 13 per cent for difficult ones. In another series of experiments he noted 7.5 per cent of correct answers for easy questions and 13 per cent for difficult questions.

The mere examination of these figures leads us towards a rational explanation of the problem. If it is supposed in fact that the horses really did possess the gift of calculation, whether this be due to their subconscious faculties, or whether they be an expression of their normal intelligence, it is difficult to explain the fact that they solve difficult problems more easily than simple ones. But if, on the other hand, the hypothesis of involuntary signals is admitted, this difficulty is removed. It can indeed be understood that the more difficult the problem set, the stronger would be the emotion of the onlookers, the more expressive their mimicry, the clearer their involuntary signals. A simple question arouses only a minimum of interest and in consequence only a feeble psychomotive reaction, whence the greater number of errors in such a case. Where the answer was not known to the experimenters, which in any case occurred only rarely, we may suppose, as we have shown in connection with the telepathy theory, that it was worked out subconsciously by one of them and then, as in ordinary circumstances, transmitted by involuntary signs.

It remains to determine the nature of these unconscious signs. We have seen earlier that Herr Oskar Pfungst believed that von Osten transmitted the required answer to his horse by means of slight unconscious movements of the head or eyes, movements which constituted a sign to stop. But this explanation does not hold good for the Elberfeld horses. In the first place all the experimenters have noted that the horses did not habitually look at the onlookers; they turned towards them only after having spelled out a word or a figure, as if they wished to find out from the spectators' attitude whether their answer contained any mistake. And in fact when there was no applause they began to spell again, and this time generally gave the right answer. Morever, in order not to be seen by the horses, Herr Krall put wide blinkers over their eyes and stood, not in front of them, but beside or behind. An attempt was made, first with Hans, then with other horses, to put a screen between the experimenters and the horse, and after a little natural hesitation the horses became accustomed to this difficulty and answered with as much accuracy as before. Experiments were made in the dark; the horses still gave the answers. Finally Herr Krall purchased a completely blind horse: Berto. The animal learned to count and calculate up to the figure 10 in a fortnight, and thereafter made rapid headway in calculating with the tens.

Given these facts we are driven to conclude that the signals supplied by the onlookers were of an *acoustic nature*. The horse has an extremely keen sense of hearing and detects sounds which are imperceptible to man.[1] When the calculating horse tapped automatically with his shod hoof there must have been among some onlookers, as he came nearer to the answer, some sort of perturbation of their breathing which ended in sudden stoppage, that is to say the cessation of chest and abdominal movements, when the number of taps corresponding to the answer was given. After that the pulmonary rhythms became normal again. The hypothesis is not simply a theory: in modern psychological laboratories it can easily be shown, by means of a recording pneumograph, that emotions do modify the breathing rhythms. Stoppage of breathing has been observed when the subject of an experiment is concentrating his thought.

In the case of contemporary "thinking" animals, and in particular those I have observed close to, the hypothesis of unconscious signals is fully borne out. Where the animal is accustomed to tap his paw against his master's palm, a slight contraction of the hand constitutes the signal to stop. I have been able to note this quite easily. In other cases there is a slight movement of the arm or shoulder which indicates to the animal that he must cease to tap or to bark. It is in this way that the dog Bim, who replied to questions by barking a specific number of times, stopped at an imperceptible involuntary lifting of the shoulder of his master, Commandant Edward Sadowski of the Polish Navy.

In the same way it is probably unconscious movements

[1] The same thing is true of many animals. My own cat Catou can hear sounds which are absolutely inaudible to normal human ears. He can also distinguish the steps of people known to him from those of strangers, even at a distance.

(apart from special circumstances dealt with later) which most often guided the Bechtereff dogs.

"My numerous assistants were always present during the experiments," the eminent Russian physiologist writes, "Dr. Vorobieva, Dr. Nikonava-Bechtereva, Dr. Chelovanoff, Dr. Flexor, M. J. Leff . . ."

With so large a number of experimenters it will be acknowledged that an unconscious signal must necessarily escape from one or another of them and so guide the dog in his actions. The signal could be an imperceptible movement of the head, or of the body or an arm, the direction of a glance, a breathing signal, etc. Moreover Professor Bechtereff indicates that the dogs had been educated by M. V. Douroff, an expert Russian dog trainer, and that *the chief experiments were carried out in his presence*. In this case the possibility of fraud comes immediately to mind, that is to say the hypothesis of a training so conducted that the dogs would obey voluntary signals, surreptitiously executed by M. Douroff.[1]

It is precisely because of special training that certain circus or music hall animals appear to carry out calculations, reply correctly to questions put to them, name playing cards, etc. The celebrated little dog Bonnie and the beautiful dog C . . . who were represented to me as having been educated by Mme. Borderieux are no more than trained animals, as I was able to see for myself.[2]

In the course of an exhibition of C . . ., her mistress

[1] In his latest book: *New Frontiers of the Mind,* Professor Rhine of Duke University, U.S.A., considers some of the experiments carried out by Bechtereff as paranormal. He also considers the "educated" horse, Lady, of Richmond, Virginia, studied by himself and his wife in 1927-1929, and by other experimenters in 1952-1953, to be capable of understanding human thoughts by telepathy. Finally he mentions, but does not describe, some recent experiments by Dr. Osiris. These consist in telepathically influencing a cat's choice of a bowl containing food. The results obtained are said to have been positive. The experiments with cats have now entered a second phase, the new tests dealing with clairvoyance.

Not knowing exactly the procedure used for the experiments with the horse Lady or with the cats it is difficult for me to judge them. It seems permissible to think that the experimenters probably did not consider the fact that the breathing rhythms of people present could furnish the animals with the necessary signals, for Rhine does not include this as a possible source of error.

[2] The female dog C . . . has been exhibited in a number of psychic circles and its right to be considered a genuine thinking and calculating animal has been acknowledged.

had the unhappy notion of appealing to me, who understood conjuring, in order to have the dog divine a "thought card;" the trick of the forced card which was employed was elementary and did not escape my notice.

Normally, dogs exhibited on the stage grasp a piece of cardboard in their jaws on which a letter, a word or a figure has been painted. The cards selected by the dog are placed side by side to make up a word or a sentence or a number.[3]

This is how the animal is trained. Before the first lesson food is withheld. Then he is allowed to sniff at a bag containing a bone or a piece of meat. The bag is thrown a certain distance; the dog goes after it to seize it in his mouth. Just when he is about to go off with it he is given a piece of sugar or a biscuit; the animal generally drops the bag in order to snap the sweet. If he does not do so, his ear is pinched, causing his jaw muscles to contract. This is repeated a number of times, the bag being thrown greater and greater distances each time, with the command "Fetch!" The animal is made to bring the bag to the trainer's feet.

Next the bag is replaced by pieces of cardboard similar to those bearing inscriptions. When the animal fetches these correctly they are arranged in a circle and by means of a leash the dog is made to go round the circle, at first accompanied. When the exercise has become habitual, he is made to go round the circle alone, the trainer staying in the center, giving "play" to the leash. It is pulled or let slip if the dog shows any desire to stray from the circular path. When he has learned how to turn round the circle of cards he is made to seize a certain card, when he reaches it, by a cluck of the tongue and the instant command to "Fetch!" Soon the tongue cluck, which can grow progressively slighter and at length absolutely inaudible to the spectators, will have the same effect as the word

[3] This is the procedure used with the Italian dog Peg, which has been the subject of enthusiastic reports by a number of journalists.

According to these accounts, Peg can add, subtract, multiply, divide, extract square roots, spell words, form sentences, distinguish colors, tell the time, etc.

For an admirer who asked how animals converse among themselves, Peg collected about twenty pieces of cardboard and aligned them methodically in order to formulate the following answer: "By moving their ears."

This, it will be agreed, is a rather unexpected answer.

"Fetch!" At this moment the instruction of the dog is complete. From then on he can answer any questions put to him. Obviously other signals can be used: different intonations of the voice when it is necessary to answer by "yes" or by "no," winking, blinking the eyes, finger movements, a slight lifting of a foot and many others.

The training of circus horse calculators is done in a similar way. The beast is brought hungry into the center of the ring, his master keeping on his left. A stable boy arrives, pours out some oats, just beyond the reach of the horse, into a little trough which he shakes continually. The animal, anxious to get away, scrapes the ground. He is then given a few oats. During the lessons which follow, the time the stable boy remains is shortened and he is brought into the ring only when the horse has been there for a certain time. The latter impatiently scrapes the ground. Finally when the reflex has been acquired the presence of the boy is no longer necessary. The animal begins his scrapings as soon as his master approaches his left shoulder where he is accustomed to be.

Next the horse is trained to stop at a signal. For this a slight cluck of the tongue is used, or a slight gesture such as a movement of the head, of the hand, or of the foot, and immediately after the animal is calmed so that he stops scraping. After a few times the signal is sufficient to cause the horse to stop. The horse is now capable of tapping the ground with his hoof a predetermined number of times and in consequence of executing calculations, or replying to questions. A radio transmitter producing "pips" in a receiver hidden in the animal's mane can also supply the desired stop signal.

Horses which write are also exhibited in circuses. According to M. Hachet-Souplet, director of the Institute of Zoological Psychology, their training is carried out in the following way:

"A rod 15 inches long, with a ring at each end, is made by a blacksmith. The right hand is passed through one of these rings, the other being held by the straps round the horse's muzzle. This second ring should be provided with sharp points on the inside, so that the slightest gesture of the hand is exactly followed by the horse's head. In this way a sort of lazy-tongs effect is obtained and it will be easily understood how in this way the horse can be made to trace out a few words. It is understood of course that when the training is complete there is no longer need to use the apparatus. The animal is now capable of writing with a brush held in his teeth."

Where performances only on the stage or in the ring are concerned use can also be made, in the training of "thinking" horses or dogs, of mechanical aids such as planks hidden in the carpet on which the animal stands and which can be raised slightly by means of a rope controlled by an assistant. The animal which puts his right hoof for example on a board of this sort, has been previously accustomed to tap as soon as the board is raised and to cease to tap when it is lowered. It wili be seen that in these conditions he can easily reply to the most varied questions put by his master or by the audience.

Let us conclude by mentioning that birds, particularly ravens, can be initiated into "calculation" by similar means. M. Pierre Gontier de Biran (known as "Barini") at one time performed an excellent "learned animal" number with his raven Toto. A music stand, holding ten pieces of cardboard bearing the numbers 1, 2, 3, 4, etc., and arranged in order one after another, was placed near the front of the stage, on the edge of a small table on which was a slate. A short distance from the table was a perch for Toto. After the usual patter, Barini took the slate and asked the public to give him a number at random. This he wrote on the slate, then, retreating towards the left and turning to the raven he asked it to fetch the corresponding card. At once the bird left its perch, seized the correct card in its beak and gave it to his master. Three or four exercises of this type seemed to demonstrate that Toto knew his figures. After this the conjuror went on to more complicated exercises. He had numbers supplied by spectators, wrote them on the slate and arranged them as additions, subtractions, multiplications, etc. The bird gave the answer required by going to the stand and picking out the necessary numbers.

In reality the bird knew no figures and could not calculate. A length of very thin thread, invisible even at very close range, was attached at one end to a corner of the slate, by the other to a sort of elastic lever fitted in the music stand and free to move in a groove. A pull on the thread displaced the lever which lifted the cards slightly. It was sufficient to hold the thread tight at the required moment for this or that card to be lifted. A spring pulled the lever back to its starting point when the thread was no longer held. The raven had been trained to seize the card which was slightly higher than the others in the series. In addition the music stand was so arranged that the animal could not take any card except the one nominated. The trick was hence foolproof. It will be realized that the conjuror worked the thread while presenting the slate to the raven.

Conclusion

My objective in writing this book can, I feel, be considered to have been attained.

First of all, I have shown that the problem of the calculating and thinking animals appears to be definitely solved. It is possible to state definitely that there are no calculating or talking animals, and this conclusion stems essentially from the fact that the psychic differences between a man and an animal are very great. Animal intelligence is primarily objective, that is to say founded upon the data of the external world, upon everything that confronts the senses and comes under their general control. Its products are the outcome of associations between old sensations and new ones and also between impressions created by these associations and preserved in the memory.

On the other hand human intelligence is at once objective and subjective. Where this second aspect is concerned it is capable of reasoning, imagining and inventing. It has the notion of "I" and the power of choice. Its intellectual operations can be almost entirely independent of sensations and of the external world.

Doubtless there is no definite hiatus, no uncrossable ditch or irreducible antagonism, between animal psychism and human psychism, since as we have seen the animal is capable of genuine reasoning in certain conditions. But in no case is his intelligence, whether conscious or subconscious, capable of carrying out intellectual operations of a complicated or abstract nature like arithmetical calculations.

And we have the right to be surprised that famous physiologists and eminent psychologists like Professor Bechtereff, Professor Ferrari, Professor Edinger and my own learned friend, Professor William Mackenzie, should have believed, even for a single instant, in the abnormal psychic faculties of "thinking animals." Their excuse must be that in their observations and experiments they did not take the involuntary psychomotor signals of respiratory origin among participants into account, and that the study of animal psychology was moreover still in its infancy at the time when they drew their conclusions. Further, they knew nothing of the methods of conjuring, which are often mingled with apparently paranormal phenomena.

In emphasizing the primordial role of these different causes of errors I hope to have put the reader on his guard against hasty and erroneous judgments if by any chance he should happen to attend any exhibitions of "transcendent" animal psychology, too often, alas, presented under the cover of psychic study.

I have the same hope if he finds himself one day in the presence of pseudo-lightning calculators, who may produce their illusions by means of certain arithmetical artifices which I have described in detail or by subterfuges which belong to the domain of conjuring.

But in contrast to what occurs in the matter of thinking animals, where only misleading interpretations or mystification are to be found, there exist genuine lightning calculators whose extraordinary faculties reveal a little of our subconscious psychism, a sort of "borderland," which can give access, even to the least daring of observers, to knowledge, acquired and admitted, of phenomena which remain provisionally listed as "paranormal."

Among many of the lightning calculators it has been noted that the gift of calculation recalls certain of the cryptic faculties of the mind studied by psychists rather than the results of a voluntary or conscious effort. This at least is what emerges from the following remarks, which I have already expressed more or less explicitly, but which I would like to sum up here. In the first place, despite its apparent relationship with the normal aptitude from mathematics, the gift reveals itself as often among people who are quite inept at mathematics, and are even unintelligent, as among genuine mathematicians. In addition, an inexplicable fact if the gift were essentiallly a conscious or normal faculty, it shows itself most often during infancy and grows less marked with age until it disappears completely. Lastly, it functions in an absolutely automatic fashion, as when the calculator solves problems while he is talking freely with other people about subjects completely foreign to the questions.

From all this it appears that the mathematical answers obtained by the lightning calculators, or more exactly by those calculators whose output is spontaneous, are the emergence in the domain of conscious ideas of a psychic operation accomplished in the depths of the being and in which the normal consciousness plays sometimes only a very small part.

And this deep region is nothing else but the *daimon* of Socrates, the *theos* of Plotin, the *planetary genius* of Paracelsus, the *transcendental ego* of Novalis, the *subliminal self* of Myers, the *unknown guest* of Maeterlinck, and, according to the term I have frequently employed, the *subconscious* or the *unconscious* of the psychists. But "unconscious," not, it would seem, because it lacks consciousness in itself, but only because our normal consciousness does not ordinarily perceive it.

In any case it appears that this cryptic "I," often unknown to consciousness, is the seat of those more or less mysterious phenomena which range from dreams to precognition, through stages which include artistic, poetical or literary creation, the intuition of genius, the gift of calculation, telepathy and extrasensory perception.

So much so that the question of the lightning calculators is integrated naturally with the general problem of the origin and significance of man's paranormal faculties. But in itself it enables us henceforward to assert, with Professor Charles Richet, that "human intelligence is far more powerful than it is thought or known to be."